SpringerBriefs in Materials

W0080020

Series Editors

Sujata K. Bhatia, University of Delaware, Newark, DE, USA

Alain Diebold, Schenectady, NY, USA

Juejun Hu, Department of Materials Science and Engineering, Massachusetts
Institute of Technology, Cambridge, MA, USA

Kannan M. Krishnan, University of Washington, Seattle, WA, USA

Dario Narducci, Department of Materials Science, University of Milano Bicocca,
Milano, Italy

Suprakas Sinha Ray ⓘ, Centre for Nanostructures Materials, Council for Scientific
and Industrial Research, Brummeria, Pretoria, South Africa

Gerhard Wilde, Altenberge, Nordrhein-Westfalen, Germany

The SpringerBriefs Series in Materials presents highly relevant, concise monographs on a wide range of topics covering fundamental advances and new applications in the field. Areas of interest include topical information on innovative, structural and functional materials and composites as well as fundamental principles, physical properties, materials theory and design.

Indexed in Scopus (2022).

SpringerBriefs present succinct summaries of cutting-edge research and practical applications across a wide spectrum of fields. Featuring compact volumes of 50 to 125 pages, the series covers a range of content from professional to academic. Typical topics might include

- A timely report of state-of-the art analytical techniques
- A bridge between new research results, as published in journal articles, and a contextual literature review
- A snapshot of a hot or emerging topic
- An in-depth case study or clinical example
- A presentation of core concepts that students must understand in order to make independent contributions

Briefs are characterized by fast, global electronic dissemination, standard publishing contracts, standardized manuscript preparation and formatting guidelines, and expedited production schedules.

Alexander Khovavko · Eugene Strativnov ·
Andrii Nebesnyi · Denis Filonenko ·
Olexiy Sviatenko · Angela Piatova ·
Maksym Barabash

Carbon Nanostructured Materials

Synthesis, Characterization, and Industrial Applications

 Springer

Authors
See next page

ISSN 2192-1091   ISSN 2192-1105 (electronic)
SpringerBriefs in Materials
ISBN 978-3-031-64120-6   ISBN 978-3-031-64121-3 (eBook)
https://doi.org/10.1007/978-3-031-64121-3

This Springer imprint is published by the registered company Springer Nature Switzerland AG
The registered company address is: Gewerbestrasse 11, 6330 Cham, Switzerland

If disposing of this product, please recycle the paper.

Authors
Alexander Khovavko
Department of Thermo-Chemical Processes
and Nanotechnology
The Gas Institute of the National Academy
of Sciences of Ukraine
Kyiv, Ukraine

Andrii Nebesnyi
Department of Thermo-Chemical Processes
and Nanotechnology
The Gas Institute of the National Academy
of Sciences of Ukraine
Kyiv, Ukraine

Olexiy Sviatenko
Department of Thermo-Chemical Processes
and Nanotechnology
The Gas Institute of the National Academy
of Sciences of Ukraine
Kyiv, Ukraine

Maksym Barabash
Department of Dignostics of Mesoscopic
Systems
Technical Centre of the National Academy
of Science of Ukraine
Kyiv, Ukraine

Department of Thermo-Chemical Process
and Nanotechnology
The Gas Institute of the National Academy
of Science of Ukraine
Kyiv, Ukraine

Department of High-Temperature Materials
and Powder Metallurgy, Department
of Foundry Production
National Technical University of Ukraine
"Igor Sikorsky Kyiv Polytechnic Institute"
Y.O. Paton Educational and Research
Institute of Materials Science and Welding
Kyiv, Ukraine

Eugene Strativnov
Department of Thermo-Chemical Processes
and Nanotechnology
The Gas Institute of the National Academy
of Sciences of Ukraine
Kyiv, Ukraine

Denis Filonenko
Department of Thermo-Chemical Processes
and Nanotechnology
The Gas Institute of the National Academy
of Sciences of Ukraine
Kyiv, Ukraine

Angela Piatova
Department of International Collaboration
National Technical University of Ukraine
"Igor Sikorsky Kyiv Polytechnic Institute"
Kyiv, Ukraine

Introduction

Carbon is an exceedingly prevalent chemical element in nature, widely employed in cutting-edge technologies. In its free state, it exhibits various allotropic modifications: graphite, diamond, and black amorphous carbon, including one of its forms, which is soot. Processes related to interfacial carbon exchange play a pivotal role in modern technologies, spanning from iron and steel production to the heat treatment of metals and alloys, fuel combustion, hydrocarbon pyrolysis, the formation of carbon nanotubes, fullerenes, and more.

Interphase carbon exchange processes encompass not only carburization or decarburization of materials but also the formation of carbides and solutions, along with the emergence of a new carbon-containing phase in the gas medium, often manifesting as soot. Investigating transformations in systems involving complex gas mixtures opens avenues for purposeful modification of atmospheric properties, thereby facilitating predictions of the physicochemical properties of the final material.

Numerous publications delve into the study of these processes, with a majority belonging to a specific group of technologies. Several studies and developments outlined in the monograph have been conducted collaboratively with the staff of the Gas Institute of the National Academy of Sciences of Ukraine, academician Borys I. Bondarenko, Viktor G. Kotov, Valerii M. Dmitriev, Olexiy P. Kozhan, Valerii S. Ryabchuk, Kostiantyn V. Simeyko, and various collaborators. The authors deem it a responsibility to extend their gratitude to all individuals involved. This publication encompasses research conducted by scientists from the Gas Institute of the National Academy of Sciences of Ukraine over the past decade, focusing on the synthesis and application of carbon materials. Specific aspects of technologies related to carbon nanotubes, thermally expanded graphite, multilayer graphene, and the production of activated carbon are elucidated.

This publication benefited from partial funding support through the Guangxi Innovation-driven Development Major Project (No. Guike AA20302013). Furthermore, partial financial backing for this publication was provided by the Project "GR4FITE3"(No. 101103752) under the Horizon Europe Program, HORIZON-CL5-2022-D2-01-01.

Kyiv, Ukraine

Alexander Khovavko
Eugene Strativnov
Andrii Nebesnyi
Denis Filonenko
Olexiy Sviatenko
Angela Piatova
Maksym Barabash

Contents

Symbols and Abbreviations

ANSYS	Software calculation program
CFD	Computational fluid dynamic
CG	Converter gas
CNM	Carbon nanomaterial
CNTs	Carbon nanotubes
CO	Carbon monoxide
CVD	Chemical vapor deposition
ETFB	Electrothermal fluidized bed
GGC	Graphite-graphene composite
GIAP-10	Catalyst for hydrocarbon conversion
KSN	Nickel catalyst
MWCNTs	Multi-walled carbon nanotubes
NPP	Nuclear power plant
OG	Oxidized graphite
PG	Producer gas
SDO	Muffle furnace
SEM	Scanning electron microscopy
SUOL	Furnace with tubular reactor
TEG	Thermoexpanded graphite
TEM	Transmission electron microscopy
WWPR	Water-water power reactor

Chapter 1
Synthesis of Carbon Nanotubes from Products of Conversion of Hydrocarbons

Abstract It presents the results of multiyear experimental and theoretical studies carried out in the Gas Institute of NAS of Ukraine on technology developing of carbon nanotubes (CNTs) manufacturing. New approaches which give the possibility to continuously receive carbon nanotubes on metal catalysts are considered. As the basic technology, it was used CVD process at moderate temperatures in the range of a kinetic-thermodynamic maximum of passing of *Bell*–Boudouard reaction. Disperse metals obtaining with pre-specified carbon value are underlaid at the basis of elaborated methodology. Products of air conversion of natural gas with strictly controllable hydrogen, carbon, and oxygen potentials were used in a role of reactionary gas from which CNTs were synthesized.

1.1 Technology of Carbon Nanotubes Production in Gas Mixtures Contained Carbon Monoxide

In the present work, new approaches were studied, which would give the possibility to produce continuous carbon nanotubes. As the basic technology, the process at moderate temperatures (in the range of a kinetic-thermodynamic maximum of passing of *Bell*–Boudoir reaction) was applied. Products of air conversion of natural gas with strictly controllable hydrogen, carbon, and oxygen potentials were used as reactionary gases. Also, the possibility of producing carbon nanotubes from a generator gas was explored. Maximum output of the final product has been achieved with an iron ore concentrate of the Inguletzky ore mining and processing plant (Krivoy Rog, Ukraine), which is used among many other materials as a catalyst for forming carbon nanotubes.

Multiyear experimental and theoretical studies carried out in the Gas Institute of NAS of Ukraine on the production of disperse metals with pre-specified carbon content are undertaken at the basis of the developed technology of carbon nanotubes manufacturing. Thermodynamic modeling of phase formation while hydrocarbons catalytic decomposition is based on software products: "GaS" and "TERRA". As result, at present, the Gas Institute has a range of the equipment allowing a deep investigation of methods to produce carbon nanotubes and catalyst for their education. The complex of equipment includes: reactors of hydrocarbon conversion, stationary and portable gas analyzers, automatic thermogravimetrical installation with controlled atmosphere, laboratory, bench, and pilot gas-thermal plants and catalysts.

Analysis of the known techniques for carbon nanotubes (CNTs) producing shows that the majority of low-temperature processes, one way or another are connected with the conversion of hydrocarbons at different temperatures [1]. In addition, these technologies differ by different catalysts and methods for their preparation. High-temperature processes for CNTs synthesis, inherently, are energy intensive and difficult to manage as well, requiring a large amount of work to clean up the primary product. Low-temperature processes, in turn, are usually inefficient and have low coefficient of hydrocarbons utilization. Developed at the Gas Institute technology of CNTs obtaining according to the authors opinion, allows to overcome the above disadvantages.

In this study, we explore new approaches, which would give the opportunity to continuously receive CNTs of uniform quality. As the basic technology, it was used process at moderate temperatures (in the range of kinetic-thermodynamic maximum of Bell–Boudoir reaction). As a reaction, gas products of air conversion of natural gas with strictly controlled hydrogen, carbon, and oxygen potentials have been used. The conversion has been carried out in the reactors of catalytic conversion using the KSN catalysts. As a result, we received the converted gas containing CO up to 18 and 35% of H_2, 3–5% of oxidants, 0.5–1.5% CH_4, and the rest N_2. The converted has been dried on silica gel and then served in a horizontal furnace on the iron ore powder, loaded into a porcelain boat and heated to the desired temperature (Fig. 1.1). First, solid-phase reduction of iron proceeds from its oxides, and then on fresh-reduced iron, which serves as the catalyst, the decomposition reaction (disproportionation) of carbon dioxide with the release of carbon materials is developing. Analysis of the carbon material on the scanning electron microscope JSM-6700F showed that there are conditions for a relatively high concentration of CNTs in the product.

The reaction $2CO \rightarrow C + CO_2$ is the main reaction for carbon delivery into the zone of nanotube formation. It should be noted that the thermodynamic possibility of carbon deposition at the presented reaction increases with temperature decreasing [2].

With temperature increasing, the kinetic conditions are improving, but the reaction is thermodynamically impossible [2]. The presence of residual methane and H_2 in methane conversion products gives the opportunity to get an additional amount of CO by the following reactions:

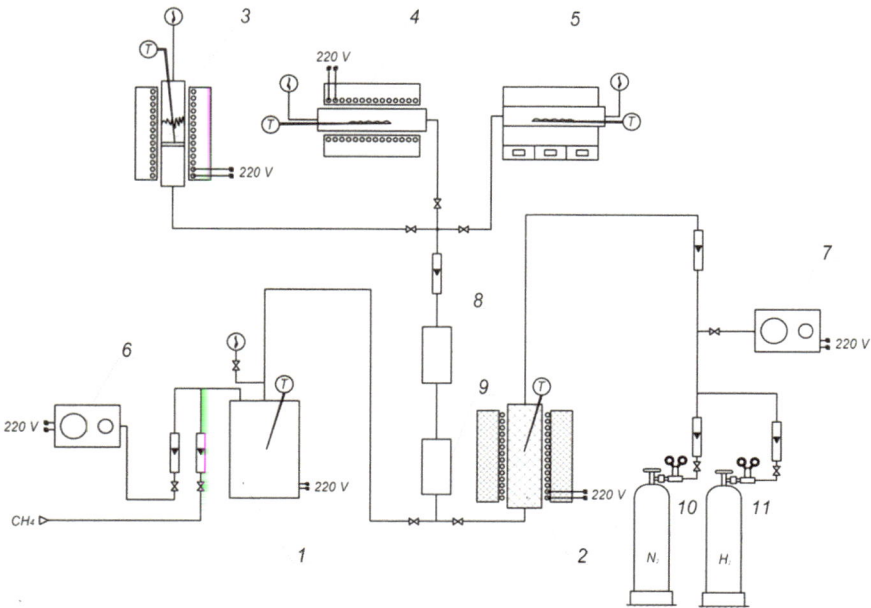

Fig. 1.1 Laboratory unit technological scheme for studying the processes of CNTs obtaining: *1*—converted gas generator; *2*—coal gasifier; *3*—vertical furnace with a fluidized bed; *4*—horizontal furnace SUOL; *5*—horizontal furnace SDO; *6, 7*—air compressor; *8, 9*—desiccant of moisture and sulfur; *10, 11*—nitrogen and hydrogen cylinders

$$CH_4 + CO_2 \rightarrow 2CO + 2H_2 \qquad (1.1)$$

$$CH_4 + H_2O \rightarrow CO + 3H_2 \qquad (1.2)$$

$$CO_2 + H_2 \rightarrow CO + H_2O \qquad (1.3)$$

Formed moisture as the result of the latter reaction, in the opinion of the authors, should gasify soot "precursors", which will ensures product uniformity. In order to more fully use the potential of converted gas, experiments were conducted in a vertical furnace with a gas filtration through a layer of iron concentrate. The formation of conventional carbon black is dominated at temperatures below 600 °C. When the temperature of the process increases, due to a sharp rise of an equilibrium concentration of carbon monoxide in the gas phase according to the reaction $2CO = CO_2 + C$ and decrease for the same reason of gas "efficiency", emission of carbon material is significantly reduced. The attempt to increase the gas flow rate resulted in the formation of the gas "channel" moving in the bed and high material carryover from the furnace. In addition, the catalyst bed began to sinter, which made difficulties with material discharge from the furnace reactor.

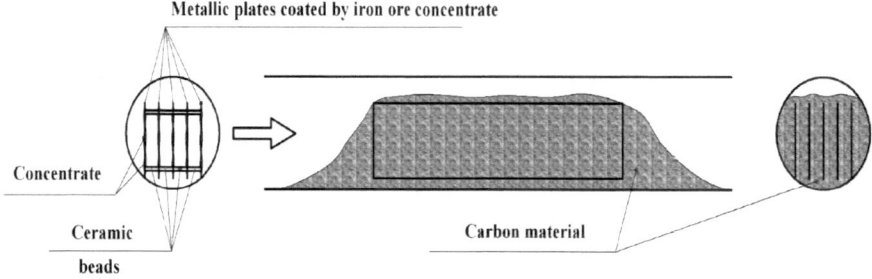

Fig. 1.2 Stack of metal plates coated with iron catalyst

To solve this problem, it was proposed to obtain the carbon material on the surface of metal plates of the stacks in such way that it uniformly filled the whole cross section of the furnace reactor (Fig. 1.2). The distance between the plates in a stack is about of 10 mm. This method reduced to almost zero gas-dynamic resistance in the furnace and increased the use of the potential of converted gas.

Also, there were studied the effects of a catalyst nature on the carbon material output. Besides the Krivoy Rog iron ore deposit there were studied nickel salts— $NiCO_3$ nNi $(OH)_2$ mH_2O, cobalt acetate, potassium dichromate—$K_2Cr_2O_7$, and copper acetate. Under equal conditions, the highest output of carbon material was obtained in the case of iron ore and nickel salt, so in the future, all studies were carried out on these catalysts.

We studied the influence of the catalyst substrate nature on an output of the carbon material. There were used different steel sorts (including stainless), as well as quartz and porcelain. The highest yield of carbon material was achieved on a carbonaceous steel and on quartz. It is noticed that the use of electrical plates (transformer) began the selection of the carbon material at temperatures above 750 °C is almost completely stopped. Electrical steels are characterized by a high content of silica; it is possible that its diffusion to the surface and subsequent oxidation contributes to decontaminating fayatilization of the surface. This leads to the formation of glassy nanofilm of Fe_2SiO_3. A similar phenomenon occurs during annealing of anisotropic electrical steel [3].

For comparison, also we explored the possibility of CNTs obtaining from the producer gas (PG). Calculations and analysis of the composition of producer gas, which was manufactured by air gasification of carbon, showed that at the gasification temperature of 900 °C, gas of equilibrium composition contains 34.1% CO and 0.4% CO_2, and at temperatures of 1000 °C 34.5% CO and 0.1% CO_2. Thus, the concentration of carbon dioxide in producer gas is almost two times higher than in the converted natural gas. This fact was the reason for parallel experiments for the process of CNTs obtaining in products of natural gas conversion and in a producer gas. Experiments have shown that, as a raw material for generating process, we have to use a material containing carbon and having a minimum content of impurities. Even in the case of wood and activated carbon used for this purpose, carbon material deposition on fresh-reduced iron occurs only if the producer gas is cleaned on the catalyst GIAP-10 (active ingredient is zinc oxide). In general, the process of CNTs growth took place sluggishly, which is apparently connected with virtually hydrogen absence in PG. Comparison experiments between CG and PG confirm notes of a number of authors about the activating role of hydrogen in the formation and growth of CNTs. Further studies were carried out mainly by using CG. Produced carbon nanotubes are shown in Fig. 1.3.

On the basis of this study, it was experimentally proved the feasibility of the developed technology. In the future, we plan to go to the furnaces of continuous action.

The use of the furnaces of this type allows to receive simultaneously an active catalyst, to move it through the furnace and keep gas potentials defined by the technology in each local zone of the furnace. Just such technology will provide continuous and stable process of CNTs production.

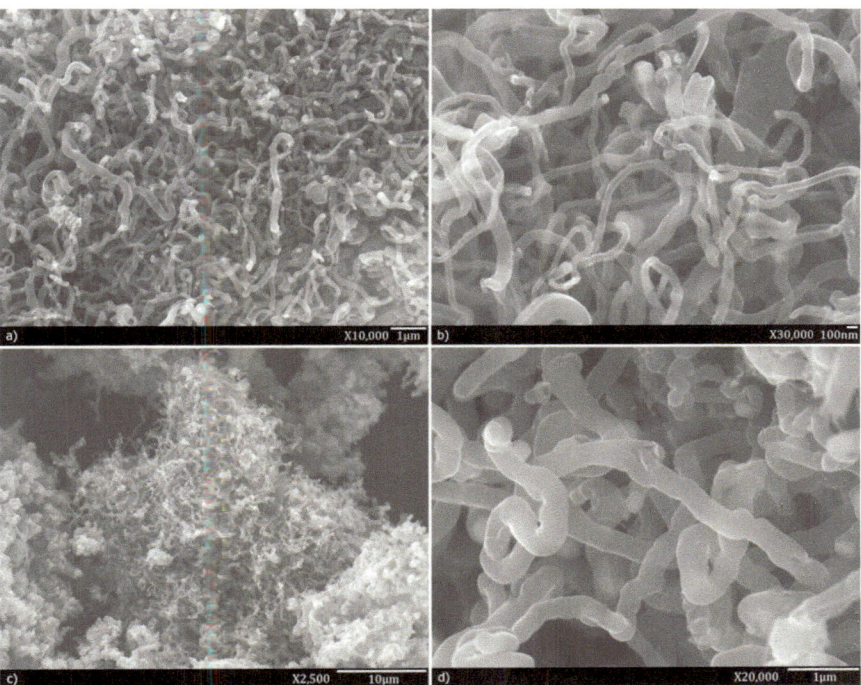

Fig. 1.3 Samples of carbon nanotubes obtained on a fresh-reduced iron: **a, b** converted gas; **c, d** producer gas

1.2 Carbon Nanotubes Synthesis from Products of Natural Gas Conversion Using as a Catalyst Fresh-Reduced Iron

The mechanism of carbon nanomaterial formation at moderate temperatures while processing of fresh-reduced iron by products of air conversion of natural gas is considered. It is shown that under given conditions, the size and the shape of the resulting carbon structures are depended on the temperature and the size of microscopic iron grains formed during reduction. These iron grains are the catalyzer of the reaction of carbon monoxide disproportion. It is concluded that the formation of nucleus of the new carbon phase occurs at the contact boundaries of neighboring grains of newly reduced iron with the subsequent formation in these places of ring-shaped carbon cuffs. Nanotubes are forming as a result of further carbon crystallization and separation of iron particles from the main mass is occurring, i.e., there is a fragmentation of the substance of the catalyst. According to the results of laboratory studies, the optimum temperature of carbon nanotubes formation in the environment of converted gas is 600–650 °C. The evidence of the hypothesis that the mechanism of the reaction of carbon monoxide disproportion flows through the intermediate stage of iron oxides formation is given.

Recirculated throat (waste) gases are typically used for cooling the product produced in shaft furnaces of direct iron reduction [4]. In these circumstances, the process of carbon formation releases. This carbon is forming from carbon monoxide contained in the waste gases. It is known [5] that the fresh-reduced iron is a catalyst in the decomposition reaction of carbon monoxide, while its decomposition in a pure gas environment at atmospheric pressure is practically not observed.

Currently, there is no generally accepted view on the mechanism of carbon monoxide formation from it, including the formation of carbon nanomaterials. Previously, it was widely believed that adsorption-catalytic theory of the dissociation lies in the basis of the mechanism of the process of carbon release from its monoxide [6, 7]. According to this theory, a significant weakening of the chemical bond between atoms of carbon and oxygen is occurred due to the adsorption of molecules of carbon monoxide on the iron surface, which upon subsequent collision of a new molecule of CO with adsorbed molecule, facilitates the destruction of this bond, according to the reaction

$$CO_g + CO_{ads.} = CO_2 + C. \qquad (1.4)$$

However, further studies showed that in chemisorption of the CO molecule on the catalyst, weakening of the bond between carbon and oxygen atoms occurs very slowly. This fact suggests that the mechanism of carbon formation is unlikely [8].

It is suggested [9] that carbon formation by decomposition of hydrocarbons can run through the stage of iron carbide formation on the catalyst, followed by its decay according to the reaction

$$Fe_3C \rightarrow 3Fe + C. \tag{1.5}$$

Carbon may be formed, depending on the conditions, in widely differing between each other flake form or in the form of carbon fullerenes [10] or even diamonds [11].

Below are results presented of research and analysis of carbon material formation on an iron catalyst in the products of air conversion of natural gas. Steel plate of Steel 3 served as a substrate for catalyst obtaining. Iron oxide (III) has been formed on these plates with the next its reduction by hydrogen in a quartz reactor. After iron reduction reactor has been blown by nitrogen. Then products of air conversion of natural gas, dried on the silica gel, were forwarded into the reactor. Products of the conversion were of the following composition (%, vol.):

$CO = 18–19$; $H_2 = 32–35$; $CO_2 = 1.5–3$; $CH_4 = 0.2–0.5$; N_2—the rest.

The study of obtained carbon material was performed by using the electron microscopes: scanning—JEOL JSM-6700F; ZEISS EVO 50 XVP and translucent. Photomicrographs of samples of material (Fig. 1.4) obtained on the iron catalyst are shown that the material consists of multi-walled carbon nanotubes (CNT_S) with outer diameter up to 300 nm. These nanotubes so called "colossal", were also obtained as a result of pyrolysis of propane-butane mixture in a plasma [12].

Based on the analysis of literature and data from laboratory studies [13, 14], the mechanism of CNT_S formation in the area of moderate temperature (up to 700 °C) in a general form can be represented as follows.

Volume of metal oxide decreases at the process of its reduction. This fact can be estimated by the expression:

$$V_{oxide}/V_{metal} = (M_{oxide}/\rho_{oxide})/(n \cdot A_{metal}/\rho_{metal}), \tag{1.6}$$

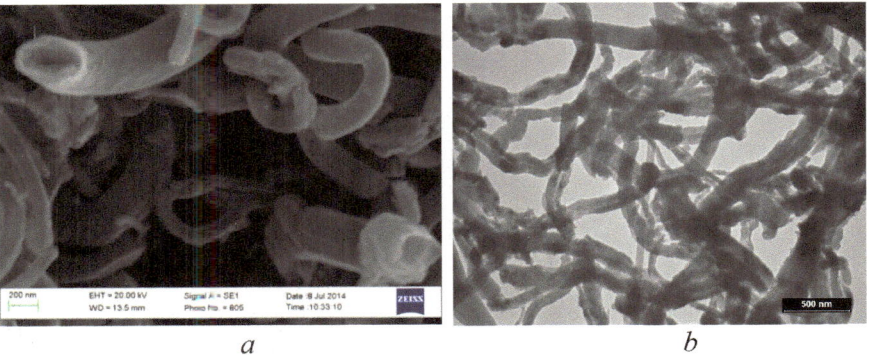

a   *b*

Fig. 1.4 Photo of the carbon material obtained as a result of fresh-reduced iron processing by converted natural gas ($T = 650$ °C)

where V_{oxide} and V_{metal}—accordingly the volumes of oxide and metal formed from it after reduction; A_{metal} and M_{oxide}—the atomic weight of the metal and the molecular weight of its oxide; ρ_{metal} and ρ_{oxide}—the density of the metal and its oxide; n—the number of metal atoms in the molecule of its oxide.

In the case of Fe_2O_3 reduction, we have: $M_{oxide} = 160$; $A_{Fe} = 56$; $n = 2$; $\rho_{Fe} = 7.9$; $\rho_{Fe2O3} = 5.1$; then after substituting these data in expression (1.6), we get:

$$V_{Fe2O3}/V_{Fe} = 2.2, \tag{1.6a}$$

i.e., the volume of iron (III) oxide at its reduction becomes smaller more than in 2 times. A highly dispersed structure of iron in the form of chaotic piles of separate grains is formed as a result of restructuring of the crystal lattice and decrease of the catalyst volume. Further, this phenomenon has a considerable influence on the formation of the resulting carbon material.

The effect of the temperature of the catalyst reduction on the size of the generating iron particles is shown in Fig. 1.5. As follows from the figure, the size of the resulting grains of iron decreases with temperature decreasing, while the mobility (self-diffusion) of atoms (ions) and, consequently, the process of collective recrystallization and consolidation of the particles slow down sharply.

As a result, the material gets a developed surface [15] and consists of small variously oriented crystals of iron, lattice of which contains many defects. Iron atoms have increased activity because of non-compensated chemical bonds, what is manifested of iron ignitibility. Boundaries are formed of the type of "insular Motta model" between touching grains of the metal [16]. Order in the arrangement of atoms could largely be absent on the boundary of grains contact according to this model. Such boundaries are called incoherent.

It is known [17] that the largest number of defects in the crystal lattice of iron is on incoherent boundaries of contact of the grains, contacting with each other. These areas have high catalytic activity. The release of carbon atoms first of all occurs at

a b

Fig. 1.5 Photo of the samples surface of the iron catalyst after reduction by hydrogen at a temperature of 450 (**a**) and 650 °C (**b**)

the process of CO decomposition with a subsequent formation of nuclei of a new solid phase. The emergence of a new phase at the grain boundaries also contributes to a higher diffusion rate of carbon atoms in these locations [18, 19]. Located on the surface of the crystal lattice of the iron atoms, in contrast to that which is situated inside due to non-compensated chemical bonds between them, a force field is created [20]. The effect of these fields is enhanced at the points of contact of contiguous iron grains. They presumably also contribute to the catalytic activity of these places and the emergence of a new phase.

Prior to carbon phase formation, the induction period exists during which there is an accumulation of free carbon atoms with the increase of their concentration up to a critical value at which the beginning of new phase formation is occurred. Study on thermogravimetric installation showed that the induction period at a temperature of 650 °C is about 8–10 min.

A new phase appearing at the points of contact of adjacent grains of iron catalyst initially has the form of a ring-shaped carbon cuff as shown in Fig. 1.6a. The similar phenomenon is observed, for example, at process of pelletizing of moist fine material, when the water cuffs are formed in places of particles contact [21]. The formation of carbon cuff is like water cuff in further contributes to the preferential formation not solid, but hollow fibers, i.e., nanotubes. The resulting carbon cuff provides for a wedging effect or grains of a metal. And as a result of a subsequent deposition (crystallization) of carbon atoms (at constant outer and inner diameters of the cuff), the separation of grains from the main "mainland" part of the catalyst is occurring (Fig. 1.6b).

Thus, the fragmentation of the grains of iron at the initial stage of the formation of carbon nanotubes occurs after dispersion of the catalyst metal at the phase of metal reduction. It should be noted that stresses occurring during carbon crystallization can reach large values. As the result cases of destruction of iron ore pellets, masonry of blast furnaces in the environment of CO-containing gases has place in practice. The atoms of carbon formed on various active centers of the surface of separated catalyst particle, due to the existing concentration gradient, diffuse toward the board of contact of the particles with the nanotube, where their crystallization is occurring.

Fig. 1.6 Mechanism of CNT growing: **a** scheme of carbon cuff formation on the contact border of the catalyst grains, **b** growth of CNTs after iron particle separation from the main part of the catalyst

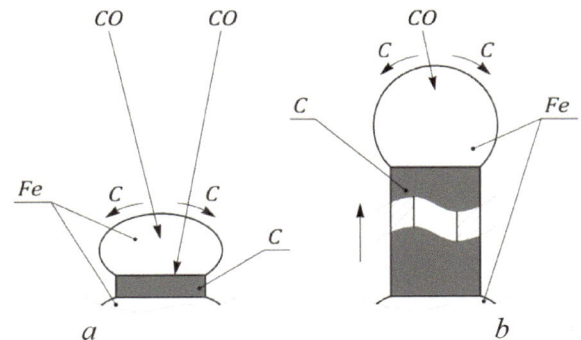

The growth of nanotube primarily oriented along the line connecting the centers of contacting catalyst grains. Because grains are mainly randomly located, that growth of nanotubes occurs in different directions relative to the wafer surface coated by a catalyst. In addition, the catalyst activity in different areas of separated iron particle is not identical, and therefore, carbon deposition rate at different points along the line of particle contact with the growing nanotube is various. This leads to curvature of the latter. Apparently, mechanical interaction of the tubes with each other at the process of their growth also contributes to curvature. As the result, the material looks like "tangled hair".

The presence of iron particles at the ends of forming nanotubes (Fig. 1.7a) may serve as indirect confirmation of the fact that boundaries of contact of the catalyst grains to each other are the initial allocation of carbon and the formation of a new phase. It is noteworthy that the particle size of the iron at the ends of the carbon nanotubes corresponds to the outer diameter of these tubes. This is clearly seen on photomicrographs (Fig. 1.7b), which shows two arrays of CNTs having a tube diameter of about 300 and 70 nm; iron particles located on their ends have the same size. Specified correspondence of the sizes allows us to conclude that the diameter of the growing CNTs is determined by the particle size of a dispersed iron catalyst, similar to the above model of CNT_S formation proposed by Baker [22].

Singled out on the active centers of the catalyst surface carbon can dissolve in iron forming an implementation of solid solution. The limiting solubility of carbon in α–iron at temperatures 600–723 °C can be estimated by the expression [7]:

$$\lg C = 3.22 - 4509/T - 2.25 \, 10^{-4}T. \qquad (1.7)$$

Here, T—temperature, K.

At a temperature of 923 K, the maximum solubility of carbon in iron is only 0.013%. Taking into the consideration the low solubility of carbon, the migration

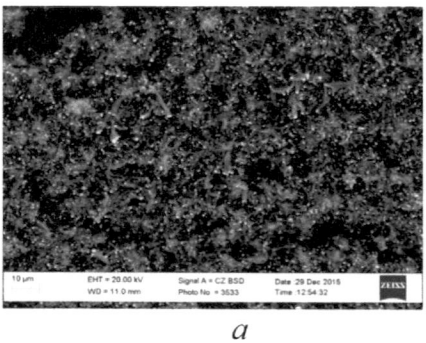
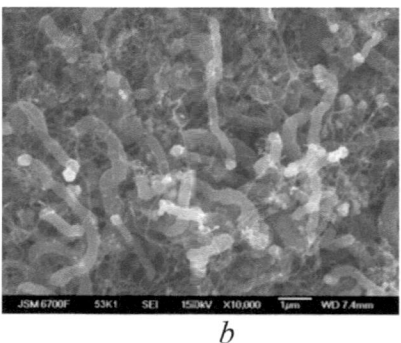

a b

Fig. 1.7 Photo of CNT_S samples illustrating: **a** fragmentation of iron catalyst, iron particles (light dots), **b** the conformity of the outer diameter of the formed CNT_S to the size of separated iron particles

of its atoms from active centers to the place of CNT_S growth, for our opinion, is mainly due to the surface diffusion, especially that its speed by 2–3 orders is higher than internal [23]. This is also evidenced by the fact that the formed nanofibers are hollow inside. Indeed, the delivery of carbon atoms into the internal part of the fibers presupposes their diffusion through the crystal lattice of iron, i.e., internal diffusion.

The formation of ferrite and cementite (Fe_3C) with polymorphic transformation of γ-Fe $\rightarrow$ α-Fe in the steel is the evidence of not sufficiently high internal diffusion capacity of carbon. Despite the fact that formation of graphite is energetically more favorable, the decomposition of austenite at low temperatures usually occurs with a pearlite formation [24], due to the relatively slow speed of diffusion of carbon atoms in iron. Apparently, carbon atoms in its crystal lattice have a mutual repulsion that is not accelerates carbon internal diffusion [25].

Studies have shown that the temperature of maximum rate of CNTs formation from products of air conversion of natural gas onto newly reduced iron is in the range of 600–650 °C. At temperatures above 700 °C, the obtained carbon material has a scaly shape, and despite the presence of iron (up to 13% or more), nanotubes are practically absent in it (Fig. 1.8a). But nanotubes are formed if obtained at this temperature carbon material to reprocess by converted natural gas already at the temperature of 650 °C (Fig. 1.8b). This fact suggests that the mechanism of CNT_S formation under these conditions proceeds through the stage of formation of iron oxides according to the reactions:

$$Fe + CO \rightarrow FeO + C, \tag{1.8}$$

$$3Fe + 4CO \rightarrow Fe_3O_4 + 4C. \tag{1.9}$$

a   b

Fig. 1.8 Photo of carbon nanomaterial obtained on the iron catalyst at the temperature of 750 (**a**) and at 650 °C (**b**) material obtained on the catalyst at 750 °C

The thermodynamic possibility of these reactions in the region of moderate temperatures can be seen from Fig. 1.6, which shows a change in their isobaric (isobaric-isothermal) potential ΔZ (Gibbs energy) depending on the temperature. Calculations of isobaric potential of reactions are performed using the data of [26]. It should be noted for some proximity in the calculation related to the lack of thermodynamic data for CNT_S formation.

As can be seen from Fig. 1.9, reactions (1.8) and (1.9) become thermodynamically impossible in the temperature region exceeding 700 °C. According to the experimental data, CNT_S formation in this field also is terminated. In the atmosphere of a converted natural gas, the iron oxides formed by the reactions (1.8) and (1.9) are reduced to the original metal:

$$FeO + CO \rightarrow Fe + CO_2, \tag{1.10}$$

$$Fe_3O_4 + 4CO \rightarrow 3Fe + 4CO_2. \tag{1.11}$$

Summing Eqs. (1.5)–(1.8), we obtain the final gross-reaction of carbon formation from carbon monoxide:

$$2CO = CO_2 + C. \tag{1.12}$$

Iron from its oxides can also be reduced in the presence of hydrogen in gas phase according to the reactions:

$$FeO + H_2 \rightarrow Fe + H_2O, \tag{1.13}$$

$$Fe_3O_4 + 4H_2 \rightarrow 3Fe + 4H_2O. \tag{1.14}$$

Fig. 1.9 Change in the isobaric potential ΔZ from the temperature of the reactions: 1—$Fe + CO \rightarrow FeO + C$, $P_{CO} = 0.18$ atm; 2—$3Fe + 4CO \rightarrow Fe_3O_4 + 4C$, $P_{CO} = 0.18$ atm; 3—$2FeO + SiO_2 \rightarrow Fe_2SiO_4$ [27]

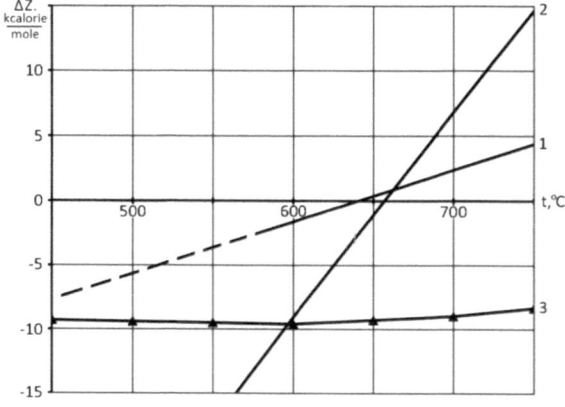

Summarizing reactions (1.5) and (1.10), and also (1.6) and (1.11), we obtain the second final gross-reaction of carbon formation occurring with hydrogen participation:

$$CO + H_2 = H_2O + C. \tag{1.15}$$

It is obvious that the higher the content of hydrogen in the source gas, the reactions (1.13) and (1.14) will run in a more degree compared with (1.10) and (1.11). This raises the proportion of carbon which is forming according to the final gross-reaction (1.15) in comparison with reaction (1.12), that was confirmed in the experiment [27].

The following experimentally observed fact is evidenced in a favor of an assumption about the possibility of iron oxides forming in the present conditions. In the case when the substrate for the catalyst is a plate of a transformer steel, the formation of the carbon material (in any form) is suppressed. It is known that the transformer steel contains high content of silica (up to 4.5% [28]). Silicon has a high affinity for oxygen, so even a slight presence in gas phase of oxidants CO_2 and H_2O leads to its oxidation [29].

Apparently, the formed silicon oxide reacts with iron oxide (II) appearing on the active centers with the formation of chemical compounds of fayalite–Fe_2SiO_4. Negative value of isobaric potential of fayalite formation (line 3, Fig. 1.9) evidences about the thermodynamic possibility of its formation, the data borrowed from source [30]. As the result, catalyst deactivation is occurred and carbon formation is not observed. Paper [31] indicates about observed in practice cases of carbon and iron oxides formation at low temperature in an environment with a high ratio of CO/CO_2 during steel annealing in heat treatment furnaces. Finally, iron oxides are detected on the X-ray pictures of iron catalyst samples after pre-reduction by hydrogen and then them processing by mixture of carbon monoxide with nitrogen [32].

Emitted carbon at temperatures above 700 °C has a scaly structure. While lower temperatures, it is synthesized in the form of nanotubes. This fact suggests that in the temperature range below 700 °C, cycles of oxidation–reduction of iron catalyst play a crucial role in the process of CNT_S formation. This factor, together with the inextricably related to the process of carbon emissions by the reactions (1.8) and (1.9), contributes to the fragmentation of iron into tiny particles smaller than 300 nm and ultimately nanotubes formation. At temperatures above 700 °C, decomposition of carbon monoxide occurs on another mechanism without the stage of iron oxides formation. In this case, although there is a fragmentation of the catalyst, the forming iron particles are larger, and emitting carbon has a scaly shape.

It is noteworthy that the visible surface of iron particles is close to spherical under the form. Special studies [33] showed that even at room temperature, crystals of various metals under the action of surface tension forces change their form in the direction of reducing their external surface. However, at low temperatures, this process proceeds very slowly and becomes noticeable only after a long time (several months). Spheroidization of iron particles proceeds much faster at temperatures of

CNT_S formation and microscopic size of iron particles. Apparently, many times repeating cycles of iron oxidation–reduction also play a big role in the process of spheroidization which leads to a weakening of the bonds between the atoms in the crystal lattice and thereby contributes for intensification of their self-diffusion.

Summary

The mechanism of occurrence and growth of CNTs on iron catalyst under conditions of low-temperature disproportionation of carbon monoxide contained in products of air conversion of natural gas is considered.

Highly dispersed structure of iron in the form of a chaotic accumulation of its particles of microscopic size is formed in terms of low-temperature reduction. The boundaries of contact of particles have the highest catalytic activity where the nucleation of a new phase in ring-shaped cuffs is occurred due to emitted carbon at carbon monoxide decomposition. Fragmentation of the catalyst with separation of iron particles from the main mass is occurring as the result of carbon subsequent crystallization. The outer diameter of the nanotubes formed while growth of carbon cuffs is determined by the size of separated iron particles. The optimum temperature for CNT_S formation from products of air conversion of natural gas is 600–650 °C. It was suggested that the mechanism of carbon nanotubes synthesis in considered conditions flows through the intermediate stage of oxidation–reduction of iron on the active centers of the catalyst.

1.3   Influence of Spent Gases Recirculation on Carbon Nanomaterial Yield Obtained from Products of Methane Air Conversion

The possibility of increasing of a specific yield of carbon nanomaterial from the products of methane air conversion through the use of spent gases recirculation is considered. The analysis of the influence of water vapor and carbon dioxide contained in the recirculated gases on the methane conversion has been performed. According to the developed calculation method, the assessment of changes in the main parameters of carbon material synthesis was done. Evaluation was performed depending on the degree of recirculation of spent gases in the transition and steady-state periods of the process. It is shown that the use of gas recirculation increases the specific yield of the carbon material, but in this case, due to the accumulation of nitrogen in the gas phase, there is a decrease in the productivity of the process for the resulting product. The evaluation of the productivity enhancing by rising of the gases pressure in the system was made. It has been established that in order to increase the specific yield of carbon material, the obtained converted gas and reusable recirculated gas should be subjected to deep purification from water vapor.

List of Symbols

æ	Degree of spent gases recirculation
V'	Quantity of gas incoming for reuse
V''	Quantity of gas withdrawn from the system
V	Total gas quantity generated in the catalytic reactor
V_s	System volume
V_{H2O}	Quantity of water vapor
V_{CO2}	Quantity of carbon dioxide
V_{CH4}	Quantity of methane
P	Pressure (atm)
T	Temperature (°C)
V'_g	Volume of gas passing through the system per unit of time
n	Recycling cycles
α	Air flow factor
ζ_i	Carbon share coming to the reactor with i component (%)
C_i	Content of CO, CO_2 and CH_4 in spent gases (%)
η_C	The carbon utilization degree of methane at formation
Π	Performance of the process of carbon nanomaterial obtaining
k	Coefficient characterizing the relative change in methane consumption
τ	Time (s)

List of Subscripts

c	Carbon
g	Gas
s	System
cr	Critical

One of the simplest technical methods for carbon nanomaterials producing is their synthesis by chemical vapor deposition (CVD) from various hydrocarbons on metal catalysts [34]. In the study [35], the possibility of multi-walled carbon nanotubes obtaining from the products of natural gas air conversion is shown. However, this process at the experimental conditions is characterized by an extremely low yield of the final product due to the low degree of carbon potential utilization of the converted gas. In this regard, the possibility of carbon material yield increasing by spent gas using due to its partial recycling has been analyzed.

1.3.1 Scheme of the Process Flow with Spent Gas Recirculation

Scheme of the process of carbon nanomaterial (CNM) obtaining with partial recycling of spent gases is presented in Fig. 1.10. Natural gas (methane), preheated air in the heater 2, and spent gas are supplied to the mixer 1 by separate streams; after mixing the mixture of gases enters the heated converter 3 to the nickel catalyst. The conversion products obtained in the converter are cooled in the refrigerator 4 and then they are dried in the adsorber 5 and sent to the catalytic reactor 6, in which the formation of the CNM occurs on the iron catalyst. Outgoing gas from the catalytic reactor is cooled in the refrigerator 7 and part of it is removed from the system through the waste valve 8, and the rest is either pre-dried in the adsorber 9 or bypassing it is sent by means of the compressor 10 to the mixer 1 and then to the converter 3.

The degree of spent gases recirculation æ is determined by the expression:

$$æ = V'/V, \tag{1.16}$$

herewith

$$V = V' + V'', \tag{1.17}$$

here V'—the quantity of gas incoming for reuse, V''—withdrawn from the system, V—total gas quantity generated in the catalytic reactor.

The composition and yield of the conversion products obtained in the converter can be calculated assuming that chemical reactions occurring at relatively high temperatures (about 900 °C) reach an equilibrium state. In this case, the difference between the actual and calculated gas compositions, as a rule, is a little [36]. It is much more difficult to calculate the performance of the process in a catalytic reactor, which

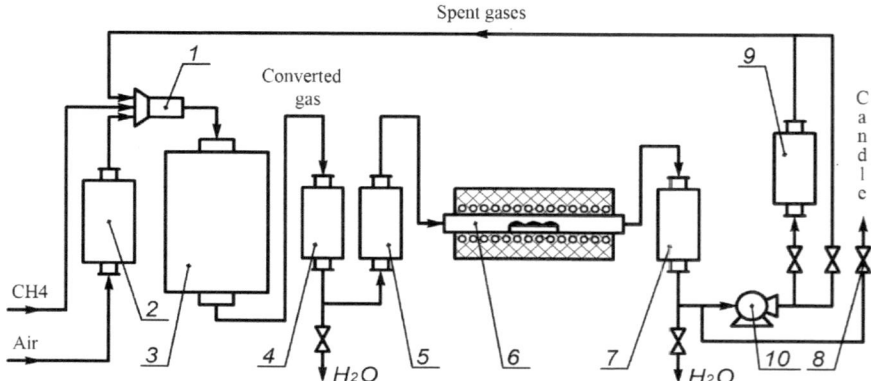

Fig. 1.10 Scheme of the process of carbon material obtaining with partial recycling of spent gases

is carried out at relatively low temperatures (about 650 °C). Because for an exact solution of the task, knowledge of the velocities of many chemical reactions, in particular, carbon formation, methane formation, water gas, is required. In this regard, the assessment of the influence of the degree of spent gases recirculation on the main process indicators (composition of the gas phase, carbon material yield, productivity) was carried out under the assumption that the chemical reactions occurring in the catalytic reactor, as well as in the converter, reach a state of thermodynamic equilibrium.

1.3.2 Initial Conditions and General Process Characteristics with Spent Gas Recirculation

The equilibrium composition of the gases generated in the converter and the catalytic reactor, as well as their volume and amount of carbon material released in the reactor, was calculated using the Gas&Solid computer program developed at the Gas Institute of the National Academy of Sciences of Ukraine [37]. The calculations were performed for the following conditions:

1. the feedstock is methane;
2. pressure in the system—1, 2 and 5 atm;
3. the temperature in the converter is 900 °C;
4. the temperature in the catalytic reactor is 650 °C;
5. gas drying 5 and 9 (Fig. 1.10) is carried out in adsorbers by silica gel.

In this case, the residual moisture content in the gas decreases to 0.01% (the dew point temperature drops to -40 °C) [38]; therefore, it was not taken into account in the calculations. It is assumed that at the process off-gases recirculation, the temperature of the gas mixture entering the converter is maintained at the same level as in the base case ($P = 1$ atm, æ $= 0$) by changing the temperature of the air preheating.

The yield of carbon material increases with increasing of carbon monoxide content in the gas; therefore, the process of hydrocarbon raw material conversion must be carried out with a minimum (close to critical) air flow rate α. A decrease of α below the critical level (α_{cr}) is fraught with the danger of soot formation, which is unacceptable due to the loss of activity and destruction of the catalyst, a sharp increase in the gas-dynamic resistance of the converter and a decrease in its performance. It has been experimentally established that when air is heated to a temperature of 500 °C, the value of α_{cr} in the base case at $P = 1$ atm is about 0.28, in this case, 2.676 mol of air is consumed per mole of CH_4.

While a transition from the base to the variant with the recirculation of spent gases, there is a transitional (initial) period during which a continuous change in the composition of the gas phase occurs and, as a result, the amount of carbon material formed in the catalytic reactor. To calculate the transition period before reaching the stationary state, when the composition of the gas phase does not change, a model of ideal displacement (the model of the "piston" mode of gas movement) [39] is adopted, in which it is assumed that the gas flow molecules move in parallel and with identical speeds. The calculation of process parameters was carried out by stages, by cycles. In the first recirculation cycle ($n = 1$), the converted gas passes through the system in an amount equal to the volume of this system. In the second cycle ($n = 2$), the same amount of gas passes through the system, but of a different composition, obtained by mixing the converted gas with the specified amount of spent gas from the first recirculation cycle. Thus, for each cycle of gas recirculation, the following expression is true:

$$n = V_g/V_s, \tag{1.18}$$

where n—the number of the natural series (1, 2, 3); V_s—system volume; V_g—the total volume of gas that has passed through the system since the transition to the process with recirculation of spent gases.

Expression (1.18) can be represented as:

$$n = k * T. \tag{1.19}$$

Here the coefficient k is equal to:

$$k = V'_g/V_s, \tag{1.20}$$

where V'_g—the volume of gas passing through the system per unit of time; T—the time since the transition to the process with recirculation of spent gases.

From the expression (1.19), it follows that under the conditions of a constant gas-dynamic mode, the number of cycles n and the time of the process with the recirculation of spent gases T are directly proportional.

1.3.3 Influence of Water Vapor and Carbon Dioxide Contained in the Recirculated Gas on the Methane Conversion Process

α_{cr} changes while the transition from the basic mode to the variant with the recirculation of spent gases due to the presence of water vapor and carbon dioxide in the last oxidizers. In this regard, there is a need to develop a method for calculating the critical value of the coefficient of supplied air to the converter for methane conversion, depending on the content of water vapor and carbon dioxide in the recirculated gas.

Figure 1.11 shows the effect of H_2O: CH_4 and CO_2: CH_4 ratio in the source gas on the possibility of carbon emission from the gas phase which reached an equilibrium state at a given temperature in the case of steam and carbon dioxide conversion of methane at a gas phase pressure of 1, 2, and 5 atm. It follows from the figure that in the case of steam reforming of methane, soot does not form at any temperature if the H_2O: CH_4 ratio in the initial mixture exceeds the value of 1.48. In this case, prima facie, there is an unexpected effect: with gas pressure increasing, the soot formation area decreases. The decrease in the critical flow rate of water vapor and the narrowing of the temperature range of soot formation boundaries in the case of a gas pressure increasing are explained by following.

It is generally accepted [40–42] that the process of steam reforming of methane can be represented by the equations of reversible chemical reactions:

$$CH_4 + H_2O = CO + 3H_2, \qquad (1.5.16)$$

$$CO + H_2O = H_2 + CO_2. \qquad (1.5.17)$$

Reaction (1.13) is accompanied by an increase in the volume of the gas phase. In this case, according to the principle of Le Chatelier, pressure increasing in the system shifts its equilibrium toward an increasing in the content of the initial, unreacted components. Therefore, the area of soot formation decreases (Fig. 1.11a) with pressure increasing and a decrease of carbon monoxide in the gas phase—the source of free carbon formation. At temperatures above 650–700 °C, carbon monoxide is already a relatively strong compound, so the upper-temperature limit of soot formation with increasing pressure practically does not change its position. The reducing of the soot formation area occurs due to the upward movement of its lower boundary.

When at temperatures less than 550–600 °C, despite pressure increasing (and the equilibrium shift of the reaction $2CO = CO_2 + C$ to the right), the degree of methane conversion by reaction (1.13) decreases so that the content of carbon monoxide in the equilibrium gas phase becomes insufficient for the formation of carbon from it.

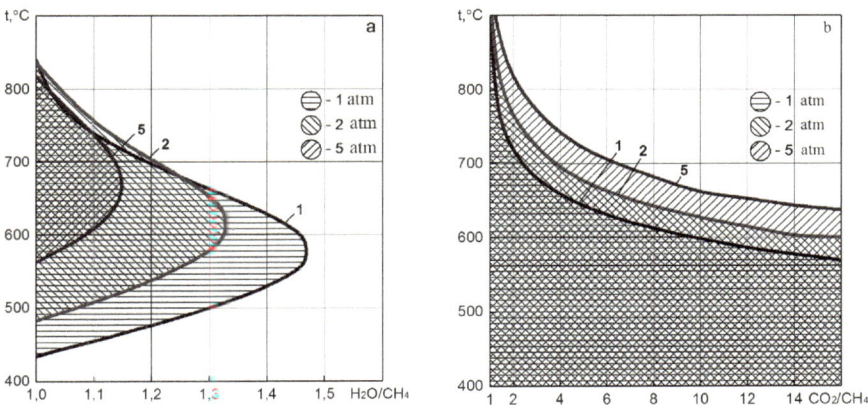

Fig. 1.11 Temperature region (hatched) of carbon emission from the gas phase, which reached an equilibrium state, depending on the ratio H_2O: **a** CH_4 and CO_2; **b** CH_4 in the initial gas at steam and carbon dioxide methane conversion; $P = 1, 2,$ and 5 atm (numbers at the curves)

Calculations show that if with H_2O: $CH_4 = 1.4$; $t = 600\,°C$; $P = 1$ atm, the equilibrium degree of methane conversion is $\eta_{CH4} = 54.7\%$, then in case of a pressure increasing under the same conditions to $P = 2$ atm, the value of η_{CH4} decreases to 41.3%, and at $P = 5$ atm to 29.1%. Thus, at $P = 1$ atm, carbon in the amount of 16.6 g per 1 m^3 of methane is released from the gas phase, and at $P = 2$ and 5 atm, carbon is no longer formed. Methane, at lower temperatures, is a relatively strong compound, and carbon is not released from it.

In the case of methane conversion by carbon dioxide in the temperature range below 600 °C, soot formation occurs even with a tenfold excess of CO_2 compared to a stoichiometric value as shown in Fig. 1.8b. The analysis shows that the effect of carbon dioxide on soot formation largely depends on the presence of hydrogen in the source gas. It is known [43, 44] that at temperatures above 700 °C, the composition of the gas phase is determined by the reaction of water gas:

$$H_2 + CO_2 = H_2O + CO. \tag{1.21}$$

Moreover, the equilibrium state of this reaction at temperatures above 800 °C is reached almost instantly [45]. In the recirculated gas, the hydrogen content can reach 30% and above, and the $H_2:CO_2$ ratio in it is ten or more. Therefore, under conditions of relatively high temperatures in the converter (900 °C), the reaction (1.18) shifts toward the conversion of carbon dioxide to the equivalent amounts of its monoxide and water vapor. Thus, under the conditions of the converter, the less active oxidizing agent CO_2 is replaced by the more active—H_2O. Obviously, this explains the proximity, found by many researches and in some cases the complete coincidence of the velocity for the catalytic conversion of methane with water vapor and carbon dioxide [46]. Consecutively summarizing the reactions (1.16) and (1.21), we obtain the final methane conversion reaction by carbon dioxide:

$$CH_4 + CO_2 = 2CO + 2H_2. \tag{1.22}$$

The total content of $H_2O + CO_2$ in the spent gases from the catalytic reactor is within 5–9%; therefore, with a not high degree of recycling, the proportion of methane, which is converted by these oxidizers, is relatively small. At high values of æ, the load on the converter over processed methane is significantly reduced; in this case, the risk of soot formation is also reduced.

Considering the above, it was assumed that water vapor and carbon dioxide in amounts of V_{H2O} and V_{CO2}, respectively, entering the converter in the composition of the recycle gas, respectively, ensure the conversion of methane without the danger of soot formation in the amount:

$$V'_{CH4} = (V_{H2O} + V_{CO2})/1.5. \tag{1.23}$$

1.3.4 Method of Calculating the Process with Spent Gases Recirculation

Based on a given degree of the recirculation, the flow of all components of the recirculated gas into the converter was determined, while its composition at first recirculation cycle corresponds to the composition of the spent gas obtained in the basic variant (æ = 0). Next, we set the methane consumption to the converter, summed it up with methane coming in with recycled gas, and the amount of methane (V'_{CH4}) that was converted by the oxidants V_{H2O} and V_{CO2} entering the converter was subtracted from the amount (determined by expression (1.23)). The remaining part of methane is converted by air, while its coefficient of discharge is the same as in the basic case, i.e., $\alpha = 0.28$ (see item 2). The consumption of methane and air in the converter was selected so that, for a given value of æ, the volumes of the gas generated in the converter in the variants with recirculation and the base one were the same. This condition arises from the requirement to ensure the constancy of the gas velocity in the catalytic reactor in the compared options for obtaining carbon nanomaterial.

Under conditions of high hydrogen content in the gas phase, the formation of carbon in the catalytic reactor from the converted gas can be represented as a joint result of the following chemical reactions according to the equations [47]:

$$2CO = CO_2 + C, \tag{1.24}$$

$$CO + H_2 = H_2O + C. \tag{1.25}$$

The presence of water vapor in the gas phase shifts the equilibrium of reaction (1.22) to the left, lowering, as a result, the yield of carbon material in the catalytic reactor. For this reason, the moisture from the converted gas was removed.

Then the composition and output of the spent gas from the catalytic reactor as well as the amount of carbon material formed in it were calculated. Having completed the calculation of the first cycle of recycling, we proceeded to the calculation of the second cycle.

Thus, using in each subsequent calculation cycle the composition of the spent gas of the previous cycle the calculation was repeated until the composition of the gas does not change, i.e., until reaching a stationary state in the system.

1.3.5 Transitional Period of the Process with Spent Gas Recirculation

If only one technological object is included in the system with gas recirculation, then the duration of the transition period is not large and can be only 3–4 recirculation cycles [48]. In this case, the system includes two apparatus of different technological purpose (converter and catalytic reactor), therefore the transition period is much longer. Table 1.1 contains such parameters: the consumption of methane and air supplied to the converter, the composition and output of the converted gas produced in it, as well as the composition and output of the spent gas from the catalytic reactor in the basic mode and after each cycle of the transition period of the process with recirculation of the dried spent gas at $æ = 0.5$; $P = 1$ atm. From the data of the table, it follows that in this case, the transition period is 8 cycles of recycling, that is the steady state of the process occurs after the passage of gas through the system in an amount exceeding the volume of the system itself by 8 times. At the same time, with an increase in the number of cycles, there is a gradual accumulation of nitrogen in the gas phase, respectively, and the content of the other components in it decreases. It is characteristic that the duration of the transition period is determined by those components of the gas whose concentration is the highest: it is, first of all, nitrogen and hydrogen. Stabilization of the concentration of the components contained in the gas in small quantities (methane and carbon dioxide) occurs much faster.

In the basic version ($æ = 0$), the specific yield of carbon nanomaterial per 1 m^3 of used methane (m_C) is 138.9 g C/m^3 CH$_4$. At the moment of transition to the process with recycling of spent gases, there is a sharp surge in the value of m_C to 225.3 g C/m^3 CH$_4$, i.e., increases on 1.6 times. With n increasing, the content of CO and H$_2$ in the converted gas, i.e., components involved in carbon formation reactions according to Eqs. (1.21) and (1.22), decreases slightly, respectively, and the yield of carbon material decreases. However, despite this, after the end of the transition period and coming on to the stationary mode, the value of m_C is significantly higher than the specific output of the CNM in the basic case. It should also be noted that the higher the degree of gas recirculation, the longer the transition period; for example, at $æ = 0.3$, it is 7 cycles, and at $æ = 0.9$, it increases to almost 30.

1.3.6 Influence of the Degree of Gas Recirculation (æ) on the Specific Yield of Carbon Material in the Steady-State Period at P = 1 atm

In the steady period of the process, with the increase of the degree of gas recirculation, as well as with the increase in the number of cycles of recirculation in the transitional period, there is an increase in the content of nitrogen in the gas phase, respectively, and the concentration in it of all other components is reduced as shown in Fig. 1.12. However, in spite of lowering the CO concentration in the converted gas, the source of CNM formation in the catalytic reactor, the specific yield of the material with æ increasing substantially rises (Fig. 1.10) as in the case of wet (curve 1) and drained (curve 2) recirculated gas using.

Table 1.1 Process indicators at the transition period, depending on the number of recirculation cycles of dried spent gases

Recycling cycles, n	0	1	2	3	4	5	6	7	8	9
Coming into the converter, mole										
Methane	1.00000	0.55680	0.54865	0.54665	0.54535	0.54530	0.54430	0.54430	0.54430	0.54430
Air	2.66672	1.48870	1.48013	1.47872	1.47636	1.47639	1.47377	1.47384	1.47375	1.47373
Spent gas	0.000	4.708	4.365	4.383	4.397	4.398	4.408	4.409	4.409	4.409
Composition of the converted gas (after the converter), % (vol.); its quantity ($V_{converted\ gas}$), mole										
H_2	37.39	36.74	36.32	36.10	35.99	35.93	35.91	35.89	35.89	35.89
H_2O	1.72	1.37	1.33	1.31	1.30	1.30	1.30	1.30	1.30	1.30
CO	18.88	17.62	17.02	16.79	16.69	16.66	16.64	16.63	16.63	16.63
CO_2	0.68	0.52	0.49	0.48	0.47	0.47	0.47	0.47	0.47	0.47
CH_4	0.04	0.05	0.05	0.04	0.04	0.04	0.04	0.04	0.04	0.04
N_2	41.29	43.71	44.79	45.28	45.50	45.59	45.64	45.66	45.67	45.67
$V_{converted\ gas}$	5.10	5.10	5.10	5.10	5.10	5.10	5.10	5.10	5.10	5.10
Composition of the spent gas, % (vol.); its quantity ($V_{spent\ gas}$), mole; the yield of the CNM (m_C), g/m³ CH_4										
H_2	31.96	31.40	31.07	30.86	30.79	30.75	30.73	30.72	30.71	30.71
H_2O	5.31	4.88	4.69	4.61	4.58	4.56	4.56	4.55	4.55	4.55
CO	10.38	9.70	9.43	9.33	9.29	9.27	9.26	9.26	9.26	9.26
CO_2	3.52	2.91	2.91	2.84	2.82	2.81	2.80	2.80	2.80	2.80
CH_4	2.41	2.33	2.28	2.25	2.24	2.23	2.23	2.23	2.23	2.23
N_2	46.39	48.60	49.61	50.07	50.28	50.36	50.41	50.43	50.44	50.44
$V_{spent\ gas}$	4.54	4.59	4.60	4.61	4.61	4.62	4.62	4.62	4.62	4.62
m_C	138.9	225.3	217.0	213.8	212.4	211.4	211.8	211.8	211.8	211.8

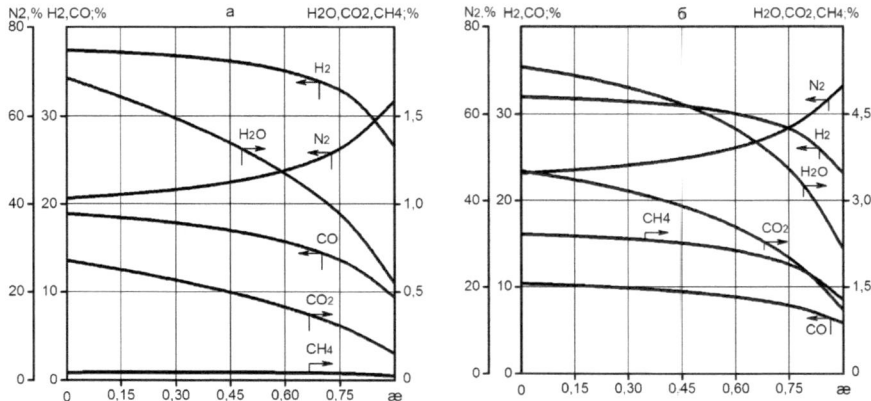

Fig. 1.12 Changes in the composition of the gas phase after the converter (**a**) and the catalytic reactor (**b**) in the steady-state period of CNM production, depending on the degree of recirculation of the dried spent gases, $P = 1$ atm

An increase in the m_C value is due to the partial utilization in the converter of carbon-containing components CO, CO_2 and CH_4 contained in the recirculated gases. The carbon share introduced into the converter by each of these components of the recirculated gas is:

$$\zeta i = 100 C_i / \sum C_i. \qquad (1.26)$$

Here, C_i is the content of CO, CO_2 and CH_4 in the spent gases, %.

The main influence on the specific yield of CNM increasing is provided by carbon monoxide: at æ $= 0.1$, the value of ζ_{CO} is 63.6%, and with æ increasing to 0.9, it increases to 70.9%. The content of carbon dioxide in the recirculated gases is approximately three times less than the CO content; therefore, its effect on the m_C is also less: at æ $= 0.1$, the value of ζ_{CO2} is 21.3%, and at æ $= 0.9$, it decreases to 13.4%. The effect of methane contained in the recirculated gases is the smallest, with æ rising, the value of ζ_{CH4} varies slightly and amounts to about 15%.

In the case of a low degree of gases recirculation of (æ < 0.5) with æ increasing, the gain in the values of m_C in the use of both non-drained and drained spent gas is practically the same. At æ > 0.5, the difference in the values of m_C in the considered variants becomes already noticeable. The lower yield of CNM in the use of non-drained recycled gas is due to the moisture presence up to 3%, which prevents a deeper conversion of CO_2 into CO, proceeding through the reaction of water gas (1.21). The higher the degree of gas recirculation, the greater the amount of moisture is returned to the converter and the greater the water vapor impedes for CO_2 regeneration. And there is the greater difference in the value of m_C when recirculating dry and wet

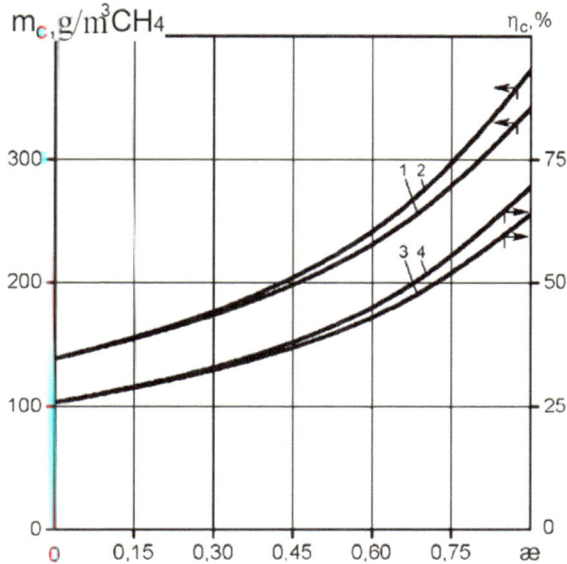

Fig. 1.13 Specific yield of carbon nanomaterial m_C (curves *1* and *2*)—not dried and dried spent gases are coming on recirculation, respectively. Also, the carbon utilization degree of consumed methane η_C in CNM forming while not dried (curve *3*) and dried (curve *4*) spent gases using in the steady-state process, depending on the degree of gas recirculation, $P = 1$ atm

exhaust gas are used in the process. For example, at æ $= 0.7$, the value of m_C is 253.5 g/m^3 CH$_4$ when non-dried recycle gas is used, and in the case of drying, it increases to 270 g/m^3 CH$_4$, i.e., increases by 6.5%. Correspondingly, with rising of the m_C values, with an increase in the degree of recirculation of non-dried and dried spent gases, the share of carbon of methane (expressed in %) coming on the formation of CNM—η_C (curves *3* and *4*) increases (Fig. 1.13).

If about 25% of carbon in methane is consumed at the base case, then, for example, at æ $= 0.75$, the value η_C increases by up to 50% in the case of non-dried recycle gas and increases by almost 55% when dried.

Thus, it is expedient to subject the spent gas supplied to the converter to preliminary dehumidification at CNM receiving with gases recirculation.

1.3.7   Effect of Gas Pressure Increasing on the Yield of Carbon Nanomaterial

The value of the critical coefficient of air flow for methane conversion changes while transition to the process of CNM obtaining with an increased gas pressure. To determine the consumption of methane and air at an increased gas pressure, the carbon absorption index (dNC) is used. The numerical value of which equals to the thermodynamically possible amount of carbon that could be gasified at converter conditions ($t = 900$ °C). In the calculation, it was assumed that the dNC rate in the basic variants at $P = 1$ atm and the increased gas pressure has the same value, equal to 0.113 mol C/mol CH_4 (calculated from experimental data) as shown in Table 1.2.

The second condition that allows calculating two unknown parameters (methane and air consumption) during the transition from $P = 1$ atm to higher gas pressure values is maintaining the constancy of the actual speed of movement of methane conversion products in the catalytic reactor. This condition arises from the requirement to ensure the constancy of the gas-dynamic regime in the implementation of the process of CNM obtaining in a catalytic reactor.

Thus, the molar flow rates of methane and air in the base case with an increased gas pressure were selected so, that the dNC and the actual amount of converted gas in the converter (taking into account the pressure correction) were the same as the base

Table 1.2 Main parameters of the process of carbon material obtaining at the steady state at varying recirculation degree of dried spent gases, $P = 2$ atm

Apparatus	Recirculation degree of spent gases	0	0.1	0.3	0.5	0.7	0.9
Converter	*Coming in (mole)*						
	Methane	2.000	1.822	1.462	1.090	0.698	0.257
	Air	5.366	4.913	4.005	3.069	2.073	0.879
	Spent gas	0	8.174	8.288	8.462	8.755	9.395
	Composition (% vol.) and the yield of conversion products (mole)						
	H_2	36.99	36.68	35.80	34.28	31.25	22.78
	H_2O	1.89	1.82	1.68	1.48	1.18	0.62
	CO	18.70	18.35	17.42	16.00	13.57	8.34
	CO_2	0.75	0.72	0.64	0.54	0.40	0.18
	CH_4	0.15	0.15	0.14	0.13	0.10	0.05
	N_2	41.52	42.28	44.32	47.57	53.50	68.03
	$V_{\text{converted gas}}$	10.20	10.20	10.20	10.20	10.20	10.20
Catalytic reactor	*Composition (% vol.), the amount of gas and released carbon (mole)*						
	H_2	28.47	28.24	27.61	26.54	24.41	18.43
	H_2O	7.26	7.09	6.64	5.96	4.82	2.53
	CO	7.96	7.84	7.51	7.01	6.17	4.28
	CO_2	4.14	4.01	3.68	3.21	2.48	1.20
	CH_4	3.83	3.77	3.60	3.33	2.81	1.61
	N_2	48.34	49.05	50.96	53.95	59.31	71.95
	$V_{\text{spent gas}}$	8.76	8.80	8.88	9.00	9.20	9.64
	C	0.60	0.59	0.54	0.48	0.38	0.19

case with $P = 1$ atm. Herewith, in the case of the transition from the basic variant to the variant with the recycling of spent gases, the influence on α_{cr} of carbon dioxide and water vapor contained in the recirculated gas was taken into account in the same way as at $P = 1$ atm.

The parameters of the process of CNM obtaining with recirculation of dried off-gases at a pressure in the system of 2 atm are presented in Table 1.2. According to the obtained calculation results, the qualitative changes in the composition of the converted and off-gases at a higher pressure in the system with an increase of æ are the same as at $P = 1$ atm. Herewith, the greater the degree of recycling of spent gases, the higher the concentration of nitrogen in the converted and spent gases and the lower the concentration of all other components.

1.3.8 Effect of Gas Pressure Increasing on the Yield of the Carbon Nanomaterial

For the starting point of reference, the performance of the resulting carbon material was chosen in the basic version of the study at $P = 1$ atm, and it can be determined by the following equation:

$$P_0 = V^0_{CH4} m^0_C, \tag{1.27a}$$

where V^0_{CH4} and m^0_C—consumption (per unit time) of methane and the specific output of the CNM in the basic version of the study, respectively.

The expression for determining the performance in the compared version (the use of gases recirculation, increasing their pressure) has a similar variant:

$$P = V_{CH4} m_C. \tag{1.27b}$$

Thus, the relative change in the performance of the process of the CNM obtaining in comparison with the base case will be:

$$\Pi / \Pi_0 = k m_C / m^0_C. \tag{1.28}$$

Here, $k = V_{CH4} / V^0_{CH4}$—coefficient characterizing the relative change in methane consumption in the changed conditions of the process compared to the base case.

Calculated by expression (1.28), the dependences of the change in the productivity of the process of CNM obtaining on going from the basic variant of operation at $P = 1$ atm to the variants with recirculation of spent gases at their different pressure are presented in Figs. 1.14 and 1.15. According to the figure, an increase in gas pressure in the system from 1 to 2 atm at æ = 0 leads to an increase in productivity by about 2.3 times; however, in the case of transition to recirculation mode, the ratio P/P_0 decreases by dependence similar linear. Productivity becomes lower than in the base case at $P = 1$ atm only when the value of æ exceeds 0.8. If the pressure in the system is increased to 5 atm, then even at æ $= 0.9$, the capacity is 2 times higher than in the compared variant.

In addition to increasing the pressure, another possible way to reduce the harmful effects of nitrogen accumulation in the gas phase during recirculation and to slow

Fig. 1.14 Specific yield of carbon material depending on the degree of spent gases recirculation at a pressure of 1, 2, and 5 atm (figures on the curves)

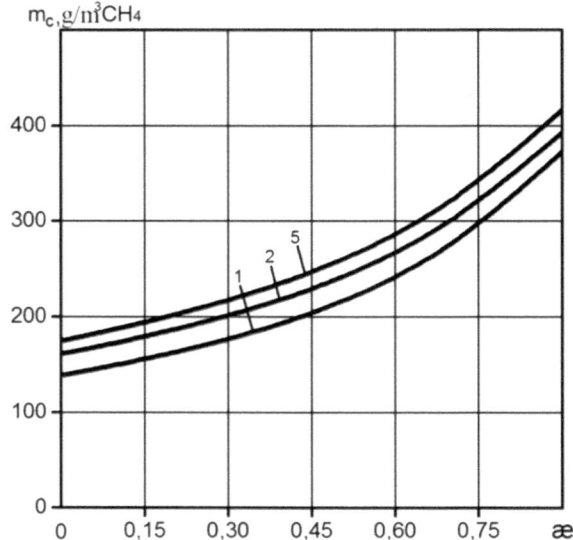

Fig. 1.15 Change in the productivity of the process of CNM obtaining relative to the base mode (æ $= 0$, $P = 1$ atm) depending on the recirculation degree of the spent gases at a pressure of $P = 1$, 2, and 5 atm (figures on the curves)

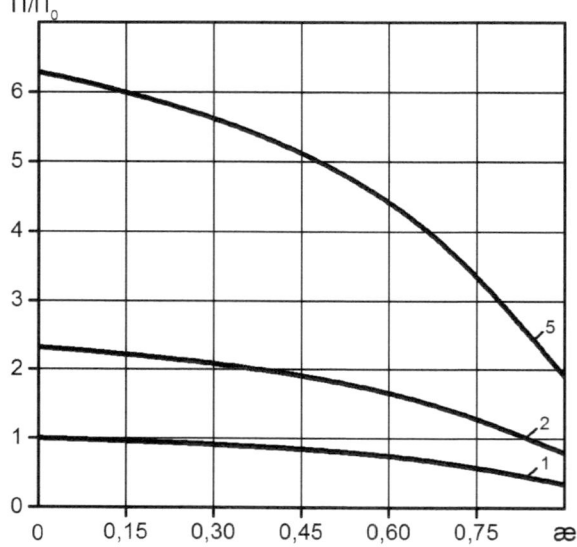

down the rate of decrease in the productivity of CNM generation process is to enrich the air supplied to the converter with technical oxygen. In this case, the conversion process can be conducted autothermally without using an external heat source to heat the converter [49].

Summary

According to the calculations, it is possible to increase the specific yield of carbon material from the products of methane air conversion by applying partial recirculation of spent gases. In this case, an increase in the degree of utilization of the carbon-forming ability of consumed methane is achieved, mainly due to the use of carbon monoxide contained in the spent gases and to a lesser extent of carbon dioxide and methane. The use of gases recirculation leads to the accumulation of nitrogen in them. As a result, the productivity of the process decreases markedly with an increase of recirculation degree. It is possible to improve the performance of the process by increasing the pressure of the gas phase in the system. In order to increase the specific yield of the carbon material, the converted and the spent gas should be cleaned of moisture.

1.4 Big Multi-walled Carbon Nanotubes Synthesis Using a Reduced Iron as a Catalyst

In the present work, there were studied new approaches which would give the possibility to continuously receive multi-walled carbon nanotubes (MWCNTs). As the basic technology, it was used a process at moderate temperatures (in the range of a kinetic-thermodynamic maximum of the passing of Bell–Boudoir reaction). Products of air conversion of natural gas with strictly controllable hydrogen, carbon, and oxygen potentials were used in a role of reactionary gas. Experiments proved the positive influence of hydrogen for MWCNTs forming. Hydrogen always presents in great quantity in the converted gas. The research of various catalysts for low-temperature MWCNTs synthesis has shown the good results of fresh-reduced iron ore concentrates. The proposed technology allows an obtaining of so-called big MWCNTs, till 300 nm in a diameter and more.

Recently, the popularity of carbon nanomaterials such as thermally expanded graphite, nanoglobals, nanowires, nanohorns, nanofibers, nanodiamonds, graphene, fullerenes, etc. is growing rapidly. Historically, it happens that scientists of Gas Institute of National Academy of Science study substances containing carbon, namely: gaseous, liquid hydrocarbons, anthracite, charcoal, coal, etc. already during the last 60 years. Back in 1968, we received a carbon material, the so-called black carbon iron [50]. Due to the objective circumstances, namely an absence of electron microscopy, capable of working at a nanoscale level, we did not even guess that these were carbon nanotubes. As it is known, for the first time, this term was introduced by the famous Japanese microscopist Iijima in 1991.

Nevertheless, we are not stopping research devoted to the production and use of carbon nanomaterials. Modern, unique installations for the production of pure pyrolytic carbon [51, 52], thermally expanded graphite [53], activated carbon [54], and carbon [13, 55] nanotubes have been created. This paper is devoted to the study of the process of multi-walled carbon nanotubes obtaining.

In order to obtain CNTs, in the majority of cases, various hydrocarbons, both liquid and gaseous, are used as raw materials. In this case, CNTs can be obtained as high-temperature synthesis (electric arc, laser) and low-temperature (catalytic) [56]. The analysis of existing methods of CNTs production demonstrates the advantages of the method of catalytic synthesis. This method is characterized by relatively low energy intensity of the process, the use of cheap raw material which contains carbon, the possibility of creating a high-performance industrial production, simplicity of equipment, and lack of thorough cleaning of the obtained product. Methods of catalytic synthesis differ one from another by the raw material from which carbon is evolved, type of catalysts on which carbon is deposited, temperature regimes. Natural gas is one of the most promising raw materials for the production of CNTs, because it is widespread, easily transported, convenient in operation, and relatively cheap. However, the use of natural gas without pre-preparation is complicated. Because, the soot which does not contain CNTs is usually formed at the direct use of natural gas. In addition, in this case, fairly high temperature is required to destroy the molecules of methane—the main component of natural gas. Even on an iron catalyst, methane begins to decompose at temperatures above 800–860 °C [57, 58]. Our studies show that in order to obtain CNTs, carbon atoms in hydrocarbon molecules are more appropriate at first to be converted into products of oxidation conversion containing CO, H_2, CO_2, and H_2O. This allows for more effective synthesis of CNTs. The article discusses the results of studies on the production of so-called big MWCNTs from products of air conversion of natural gas—distributed hydrocarbon fuels.

It is known that while heating a gas phase containing CO, the formation of solid carbonaceous material occurs due to the process of disproportionation (decomposition) of carbon monoxide by the reaction:

$$CO = CO_2 + C. \tag{1.29}$$

The catalysts of this reaction are reduced iron, nickel, cobalt, and palladium [59]. Studies have shown that freshly reduced iron gives the best results. Therefore, the experiments started on iron ore concentrate in its usual (oxidized) state.

Typically, converted natural gas contains free hydrogen, whose content is almost two times greater than CO and reaches 35%. Therefore, it is interesting to find out the role of hydrogen in the process of carbon formation from its monoxide. The thermodynamic analysis of carbon formation from mixtures of CO + H_2 was made using the computer program "GaS", which was developed at the Gas Institute of the National Academy of Sciences of Ukraine [60]. The analysis of thermodynamic calculations led us to the conclusion that the formation of carbon in the presence of hydrogen is due to the exothermic reaction:

$$CO + H_2 = C + H_2O \qquad (1.30)$$

Considering that for the formation of the same amount of carbon by reaction (1.30), it takes two times less of carbon monoxide than by reaction (1.29), and the initial addition of hydrogen leads to a significant increase in the thermodynamic possibility of carbon formation. For example, if at a temperature of 650 °C 196.4 g of carbon is formed from 1 m^3 of pure CO, then from the gas consisting of 1 m^3 of CO and 0.43 m^3 of H$_2$ (the mixture contains 30% hydrogen), more carbon is formed—213.5 g. The location of the maximum of carbon yields slightly depends on the temperature of the process and is in the range of concentrations of H$_2$ from 33 to 40%. If the content of hydrogen in gas is more than 40%, there is a shortage of CO—the source of carbon, as a result, the possibility of carbon formation sharply decreases. The conclusions of the thermodynamic analysis are confirmed by our experimental data, some of which are shown below.

A scheme of the laboratory installation for MCNTs synthesis from products of air conversion of natural gas is shown in Fig. 1.16. Natural gas was subjected to air conversion in a catalytic reactor 1 located in a heating electric furnace 2. The catalytic reactor was filled with a nickel catalyst GIAP-3-6H. Before entering the reactor, natural gas was blended with air, which was preheated to a temperature of about 550 °C in an electric air heater 3; the air flow rate was about 0.3. The consumption of natural gas and air was measured using flowmeters 4. The temperature in the catalytic reactor 1 was maintained at about 950 °C. The resulting converted gas at first was cooled in a refrigerator 5 and then dried in a moisture absorber 6 filled with silica gel. The dried gas was fed to the reactor 7, which is heated by the electric furnace 8 (horizontal or vertical).

Various iron ores and concentrates, salts of iron, nickel, and others as catalysts are put in the reactor 7. Substances were loaded in porcelain boats, mesh gas-permeable baskets or applied to a thin layer (up to 1 mm) on a metal plate. Subsequently, after the converter 1 run to the mode and heating the reactor 5 to the desired temperature, nitrogen was fed into the reactor. Then hydrogen was supplied to the reactor to reduce iron (nickel) from their oxides.

During the experiment, gas samples were taken for chemical analysis before and after the reactor 5, which was analyzed on the gas chromatograph 6891N (Agilent, USA). Analysis of samples of carbon material was carried out on an electron microscope ISM-6700F. An elemental analysis was carried out on a ZEISS EVO 50 raster electron microscope.

Initially, a study was conducted to determine the temperature at which the process of carbon formation occurs at maximum speed. This temperature was taken as the "base" from which it began to evaluate the efficiency of carbon material obtaining while selecting the method of preparing the catalyst itself.

The dependence of carbon content in the iron catalyst on the temperature of the process is shown in Fig. 1.17. From the diagram, it follows that at the experimental conditions, the maximum rate of carbon emission was observed at a temperature of about 650 °C. At lower temperatures, the process of carbon formation is inhibited by kinetic reasons due to the sharp drop in the rate of chemical reactions. As the

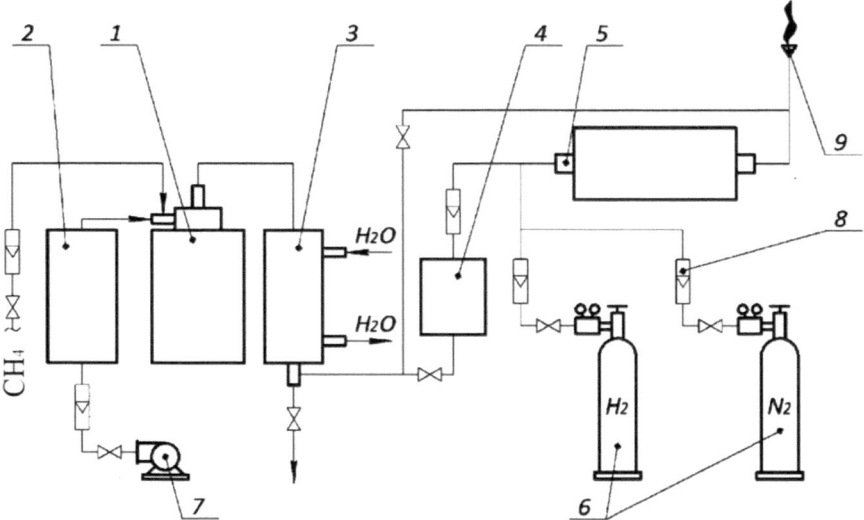

Fig. 1.16 Scheme of the experimental installation: *1*—converter; *2*—air heater; *3*—cooler; *4*—moisture absorber; *5*—reactor; *6*—cylinders with hydrogen and nitrogen; *7*—compressor; *8*—flowmeters; *9*—burning candle

temperature rises, the rate of chemical reaction increases sharply, but at $t > 650\,°C$, a carbon monoxide deficiency may occur, which may participate in carbon formation reactions.

It should be noted that the experiments did not establish a noticeable effect of the residual methane in the conversion products on generated carbon. Also, the influence of the nature of the catalyst substance on the yield of carbon material was studied. In addition to the iron ore concentrate of the Kryvyi Rih (Ukraine) deposit, nickel salts—$NiCO_3 \cdot nNi(OH)_2 \cdot mH_2O$ and iron—$Fe(NO_3)_3 \cdot nH_2O$, cobalt acetate, potassium bichromate—$K_2Cr_2O_7$, copper acetate were investigated. At equal conditions, the largest yield of carbon material was obtained in the case of the use of iron ore concentrate, salts of iron and nickel.

The influence of the nature of the catalyst substrate on the yield of carbon material was studied. Different steel grades including stainless, as well as quartz and porcelain, were used. The largest yield of the carbon material is achieved on steel st.3 and on quartz. It is noted that in the case of the use of plates from electrical (transformer) steel, the carbon content is not significant, and at temperatures above $750\,°C$ ceased practically completely. Electrical steels are characterized by high content of silica; it is possible that its diffusion on the surface and subsequent oxidation contributes to deactivating of the surface. That is, it leads to the formation of a glass-like nanofilm Fe_2SiO_3. A similar phenomenon occurs at an annealing of anisotropic electrical steels [3].

Fig. 1.17 Temperature dependence of carbon material yield on reduced iron

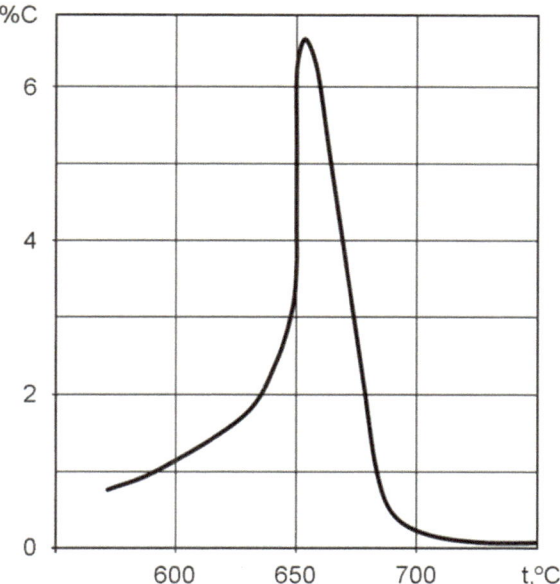

Taking into account that the concentration of carbon monoxide was twice as high in a generator gas as in a converted natural gas, the possibility of CNTs obtaining from a generator gas was studied. The calculations and analysis of the composition of generator gas obtained by air charcoal gasification have shown that at gasification temperature of 900 °C, the gas of equilibrium composition contains—34.1% CO and 0.4% CO_2, and at temperature 1000 °C—34.5% CO and 0.1% CO_2.

The results showed that even in the case of wood or activated coal used for this purpose, forming of the goal carbon material occurs only at purifying of the resulting generator gas from sulfur on the GIAP-10 catalyst (active element—zinc oxide). In general, the growth of carbon nanotubes proceeded sluggishly, which may be explained by the lack of hydrogen in the generator gas. A comparative analysis of two cases confirms the idea of many authors about the activating role of hydrogen in the process of CNTs synthesis. The following experiments were conducted only on converted gas. Taking into account the fact that the raw material for the preparation of the iron catalyst is readily available and cheap, CNTs obtaining was carried out on the iron catalyst.

From the obtained carbon material, samples were taken which were examined under an electron microscope. The increase of 1000 times is insufficient, because it does not allow to clearly examine the structure of the material (Fig. 1.18a). The increase should be (10–50) 10^3 times; in this case, carbon nanofibers (nanotubes) are well visible (Fig. 1.18b, c, e). The photo shows that the carbon in the material is almost entirely in the form of fibers and is not contaminated with a soot. However, it contains a certain amount of iron, which is in the form of small particles close to a spherical shape in diameter up to 200–300 nm. Analysis of the material on a raster electron microscope showed that the average content of iron in the material can reach 2–3%.

Particles of iron are mainly on the upper rising ends of nanofibers. Most of the fibers (tubes) have approximately the same diameter as the diameter of the iron particles, that is, up to 300 nm. However, in some cases, in the main array, there

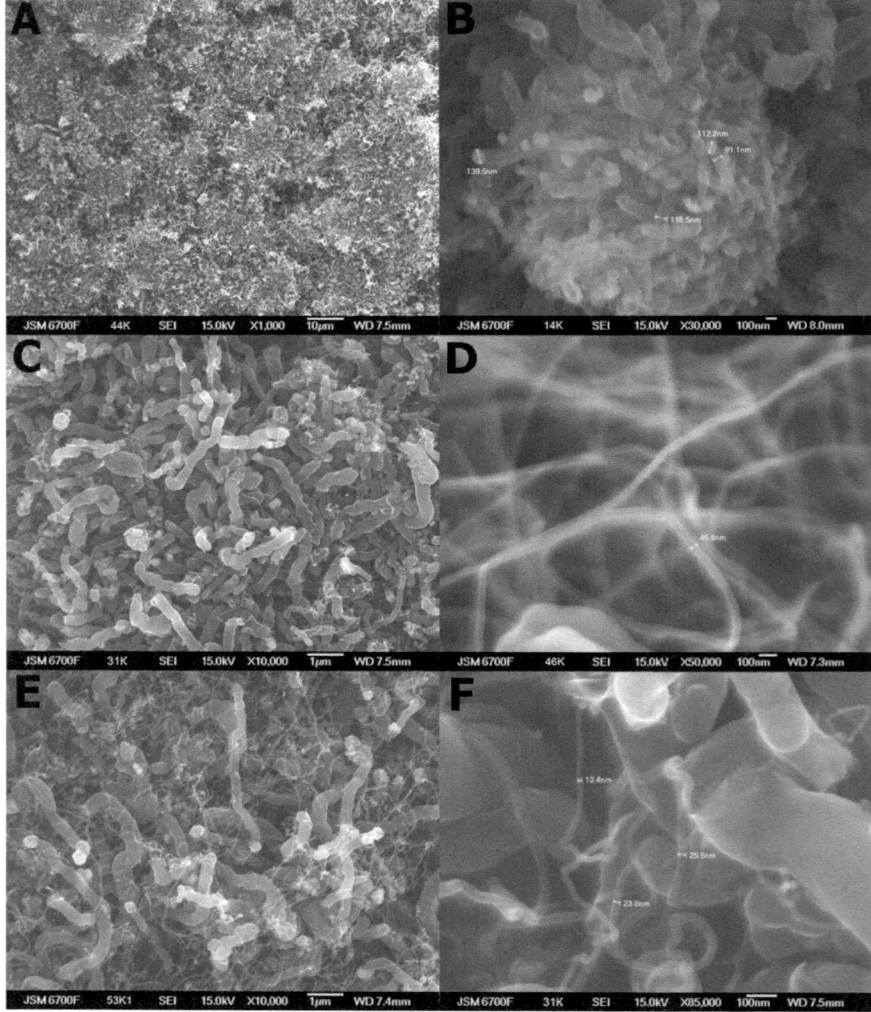

Fig. 1.18 General view of produced MWCNTs (scanning electron microscope—ISM-6700F). Increasing in times: **a** 1000; **b** 30,000; **c** 10,000; **d** 50,000; **e** 10,000; **f** 10,000

are more thin fibers with a diameter of about 20–50 nm (Fig. 1.18d, f). Perhaps the diameter of the fibers is determined by the size of the iron particles that are turn off the surface of the iron catalyst in the initial moment of the process of carbon formation.

The feature of the proposed technology is the simplicity of obtaining so-called large nanotubes. These are nanotubes with a diameter of 300 and more nm. The basic principles and so the main advantages of our method are: moderate temperature, atmospheric pressure, using of gas mixes containing CO—products of air-gas conversion of non-expensive component—natural gas. Figure 1.19 demonstrates the TEM photo of obtained MWCNTs; in turn, Fig. 1.20 demonstrates Raman spectra of synthesized MWCNTs.

The purity of the resulting nanotubes reaches 95% by carbon. It should be noted that our nanotubes do not contain a traditional harmful impurity as an amorphous carbon—soot. Such nanotubes are now regarded as the most promising material for

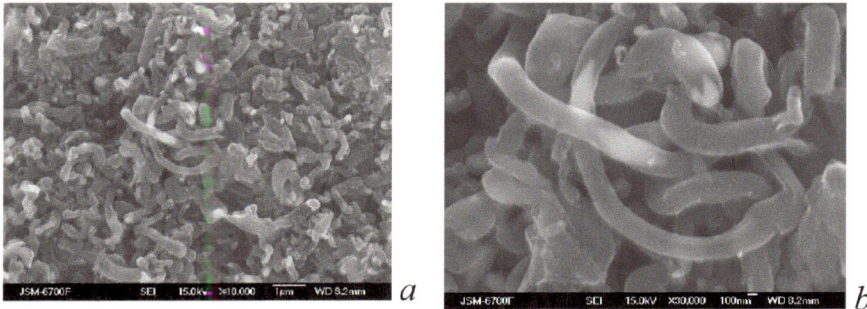

Fig. 1.19 SEM photo of big MWCNTs obtained at different temperature modes and gas composition: **a** SEI enlargement 10,000, **b** SEI enlargement 30,000

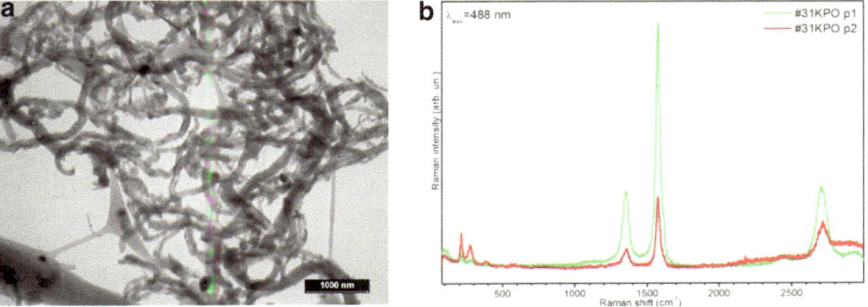

Fig. 1.20 Morphology of MCNTs and Raman spectra: **a** TEM photo of obtained MWCNTs, **b** Raman spectra of MWCNTs

medicine, electronics, and even ultralight body armor. Calculations of the technological parameters for the manufacturing of 1 ton/day were made to realize a large-scale production of MWCNTs by developed method.

We have a wide range of opportunities to reduce prices and increase the productivity of equipment by creating modular equipment and using instead of natural gas a propane-butane mixture with a much higher carbon content and using exhaust gas for heating instead of electricity.

Summary

The technology of big MWCNTs synthesis of high purity, based on converted hydrocarbons using, namely CH_4, has been proposed and explored. Only the positive influence of hydrogen, which is always present in a large quantity in the products of methane conversion, is shown. Various catalysts of low-temperature synthesis of MWCNTs have been investigated, and it is shown that freshly reduced iron ore concentrates are rather good for the offered method. The proposed technology allows receiving so-called big nanotubes, which are one of the most promising carbon nanomaterials. In the future, it is planned to obtain the MWCNTs in continuous furnaces.

1.5 Multi-walled Carbon Nanotubes Synthesis on Iron Ore Pellets by CVD Method

The aim of this work was to determine the efficiency of iron ore pellets used as catalyst for carbon nanotubes (CNTs) growing by chemical vapor deposition (CVD). The products of methane air conversion were used as a working atmosphere and source of carbon. Our previous experimental studies and thermodynamic calculations made it possible to choose the optimal temperature of the process. The choice of a catalyst was dictated by the idea of using ready-made materials that do not require preliminary preparation. The performed studies have shown both the high efficiency of CNTs growth on iron catalysts and the high and uniform quality of the resulting product. Raman spectra confirmed the presence of CNTs with two characteristic peaks at 1310 and 1578 cm^{-1} corresponding to D and G modes, respectively. Obtained CNTs by CVD method are multi-walled and most likely have a curved structure.

Research was conducted with the aim of studying the possibility of integrating CNTs production into the real metallurgical processes, for example, direct reduction of iron, the technological atmospheres of which mainly consist of a mixture of H_2 and CO. The Gas Institute of the National Academy of Sciences works in the field of direct production of iron in complex gas systems. Based on many years of research experience recovery of iron ore raw materials, we made an attempt to combine the process of obtaining metallized iron—a catalyst with the simultaneous synthesis of CNTs. Similar studies were carried out in the work [61].

The method of carbon nanotubes (CNTs) synthesis by chemical vapor deposition (CVD) is currently modern and widely used in industry [62–64]. The advantage of the CVD method is its simplicity and undemanding technology, an available carbon precursor-carbon-containing gases, and CNTs yield up to 95%. The possibility of sustainable growth of CNTs on surfaces of various shapes and sizes with low energy consumption is essential important factors affecting the synthesis of CNTs by the CVD method are: the choice of precursor gas, temperature, and pressure in the reaction chamber, gas flow rate, nature, and size of the catalyst grain. Hydrocarbons are usually used as carbon precursors: methane, acetylene, ethanol, ethylene, cyclohexane, propylene, benzene, etc. Metals of the iron subgroup Fe, Co, Ni, and their mixtures are effective catalysts for the formation of CNTs [65]. SiO_2 and Al_2O_3 are usually used as catalyst carriers.

Despite the variability of the methods of CNTs' synthesis [65, 66] which are used in various fields of science and technology [67–69], it becomes clear that such an interdisciplinary nature of processes description occurring in nature is not accidental; it reflects the fundamental principle of the material unity of the world, which is based on universality and the absoluteness of matter that surrounds us.

Another important scientific task today is the study of the physical and technological properties of various CNTs, and the potential of their application will determine the further progress of mankind. In this context, the unification of this topic of CNTs obtaining and researching, establishing key technological parameters that affect their architecture becomes urgent. The nature of the catalyst, the ratio of its components, and the method of CNTs' synthesis are the steps in solving the mentioned problem and have a significant impact on CNTs' type, structure, growth rate, diameter, and chirality.

1.5.1 Scheme of Thermogravimetric Unit for the Study of Kinetics of Carbon Nanomaterial Growth

The process of CNTs' synthesis was carried out using one of the most affordable species of hydrocarbon gases–natural gas as raw material. The main component of natural gas is methane, the content of which can reach 90% or more. The connection in the CH_4 molecule between carbon and hydrogen atoms is relatively strong; therefore, even using the catalyst, it requires increased temperatures to decompose methane, which creates prerequisites for the formation of amorphous carbon together with nanotubes. In order to transfer carbon into the CH_4 molecule into a less strong connection, natural gas is previously converted, turning the methane into an unstable at low-temperature carbon monoxide (CO). The use of products of natural gas conversion made it possible to reduce both the synthesis of the carbon nanotubes to 873–923 K, and the content of amorphous carbon in them [70].

For natural gas converting, a converter created at the Gas Institute of National Academy of Sciences was used (Fig. 1.21).

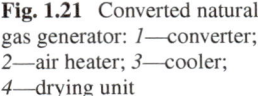

Fig. 1.21 Converted natural gas generator: *1*—converter; *2*—air heater; *3*—cooler; *4*—drying unit

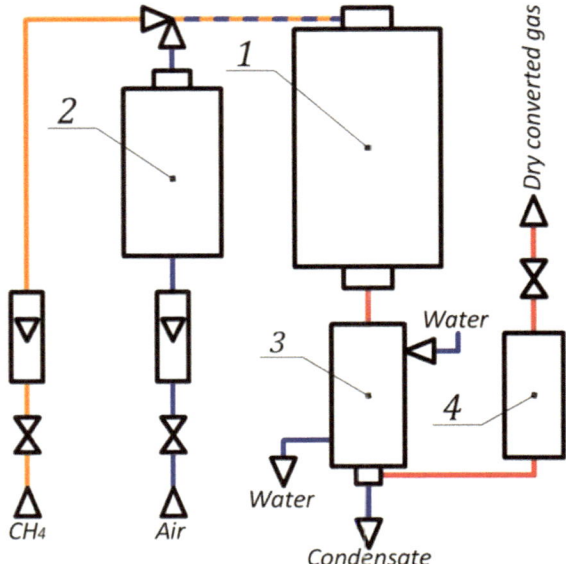

The converter was heated by an electric furnace; the temperature in the catalyst layer was about 1223 K. The conversion of natural gas was carried out with air heated to a temperature of 823 K in air heater 2; the air flow rate was 0.27–0.3. The resulting converted gas was first chilled in a cooler 3 and then dried in a moisture absorber 4 filled with silica gel. As a result, converted gas containing up to 19% CO and 35% H_2, 1.5–2% CO_2, 0.5–1.5% CH_4, and the rest N_2 was obtained. Dried converted gas was supplied to the kinetic thermogravimetric installation for the study of the process of CNTs obtaining (Fig. 1.22).

R&D to obtain CNTs by the CVD method has been carried out for the last decade at the Gas Institute of the National Academy of Sciences of Ukraine. A multifunctional laboratory complex was created for this purpose. The equipment makes it possible to obtain complex gas atmospheres with the possibility of varying their oxygen-carbon potential. The complex includes three-zone horizontal and vertical furnaces with precise control of the heating rate and exposure of the sample in a given atmosphere, gravimetric unit for the study of kinetics of carbon nanomaterial growth [71–74].

The process of carbon nanomaterial growth took place on the prepared iron ore pellets in the quartz reactor of the thermogravimetric kinetic unit (Fig. 1.22). In our previous work [74], it was determined that the highest rate of carbon yield on iron catalyst occurs at a temperature of about 923 K, so the reactor was previously warmed up to this temperature.

Heating to a given temperature was performed in an atmosphere of nitrogen, after which nitrogen was replaced by converted gas. The consumption of converted gas was about 2 l/min. The process of carbon nanomaterial deposition was controlled by the growth of the mass of the sample. After the end of the process, the reduction

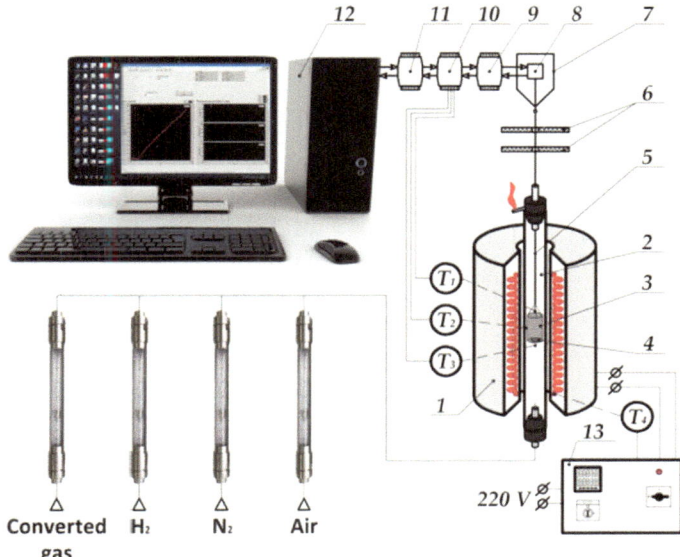

Fig. 1.22 Scheme of thermogravimetric unit for the study of kinetics of carbon nanomaterial growth: *1*—electric furnace; *2*—reactor; *3*—mesh basket; *4*—sample; *5*—wire; *6*—heat shields; *7*—balance; *8*—strain gauge load cell; *9*—analog signal input module from the strain gauge; *10*—analog signal input module from thermocouples; *11*—analog-to-digital converter; *12*—computer; *13*—control unit; T_1–T_4—thermocouples

atmosphere was replaced by an inert atmosphere-nitrogen in which the samples were cooled to room temperature. Cooled samples were cleaned of carbon, and the mass of the received material was determined. Morphology of synthesized nanomaterials was studied using methods of scanning electron microscopy (SEM) using the JSM-6490LV microscope.

Laboratory experiment. Laboratory experiment has been carried out to study the kinetics of carbon deposition on iron ore pellets (Fig. 1.23). Composition of roasted pellets, % wt.: $Fe_{total} = 68.0$; $Fe_2O_3 = 96.51$; $FeO = 0.58$; $SiO_2 = 2.19$; $Al_2O_3 = 0.41$; $CaO = 0.16$; $MgO = 0.105$; $S = 0.02$; $P = 0.025$. The studies were carried out on a gravimetric kinetic setup. In the course of the work, the weight gain of the sample was determined in comparison with the initial value depending on the time of the process and at a constant temperature. At first, it was a loss of mass associated with the reduction of iron oxides included in the pellets after the supply of natural gas conversion products.

This mass loss was not taken into account when plotting the curves. After the completion of the reduction process, the process of carbon deposition began on the surface of the samples. The experimental dependences of the weight gain of the samples from the heating time are shown in Fig. 1.24.

Fig. 1.23 Iron ore
pellets—the initial material
for carbon nanomaterial
growth

Fig. 1.24 Kinetic growth of
CNTs on fresh-reduced iron
pellets

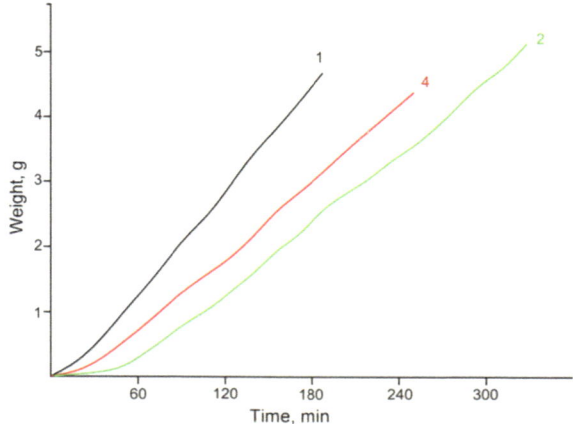

Based on the obtained experimental data, presented in the graph, it is possible
to draw a conclusion about a high increase in the mass of deposited carbon in the
form of CNTs. Thus, it can be concluded that the proposed technology is effective
for obtaining this kind of materials.

In all experiments, a high increase in the mass of carbon material (5 g and more)
was observed. A characteristic feature of the experiments was the fact that we did
not manage to fix the end of the growth of carbon material. Due to the fact that the
deposited material did not fit in the volume of the mesh basket and began to fall out
of it, after which the experiment was stopped.

Analysis of the averaged, kinetic, experimental data showed a rather high efficiency of the process of carbon nanomaterial growth. A carbon material was obtained
in the form of multi-walled CNTs, which can be judged from the data of electron microscopy (Fig. 1.25) and Raman spectroscopy. The increase in the mass of
deposited CNTs exceeded the mass of the initial samples of iron ore pellets by more
than five times.

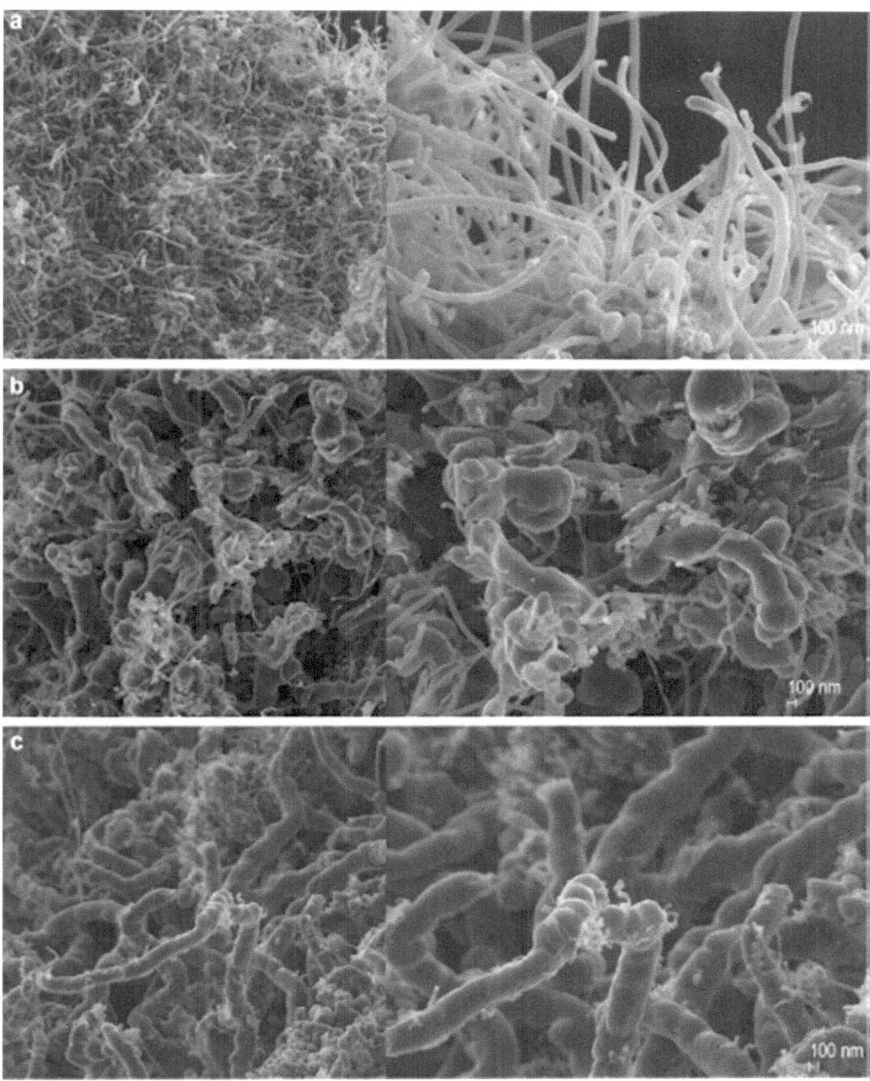

Fig. 1.25 SEM photo of synthesized CNTs on fresh-reduced iron pellets: **a** sample 1; **b** sample 2; **c** sample 4

1.5.2 Raman Scattering Spectra of Carbon Nanotubes Obtained at Different Synthesis Regimes

Raman scattering spectra of four samples were studied using a mini-Raman Pro spectrometer manufactured by Lightnovo (Denmark). A laser with a wavelength of $\lambda = 785$ nm for the spectrum region of 600–2000 cm^{-1} was used for shooting. The accuracy of determining the wave number belongs to the range of 0.5–2 cm^{-1}. Registration parameters: exposure—100 ms, 50 spectra were averaged. In each case, the Raman spectra of all samples were recorded using the same shooting parameters. The powder of carbon nanomaterials was placed on metal substrates, which are inert to the laser that was used to capture the Raman spectra. The first sample was recorded at a laser power of 86 mW. The second sample was recorded with the same parameters and under the same conditions, but during the recording, the powder began to burn. Burning was characteristic of the third and fourth samples as well. In the second, third, and fourth samples, the laser power was reduced and the spectra were obtained at a power of 50 mW. At recording the Raman of the third sample, the spectrum was not similar to the spectrum of normal nanotubes. This appearance of the Raman spectrum is not inherent for the normal spectra of characteristic modes of nanotubes.

Raman spectroscopy is a technique of fast and non-destructive control of electronic and phonon parameters of carbon materials, which are correlated with their chemical composition and structure at the level of chemical interaction. In Raman spectroscopy spectra (Fig. 1.26a–c) of the obtained materials, the position, half-width, and relative intensity of the spectral bands, which are determined by the carbon structure, are characteristic. Usually, the Raman spectra of multilayer carbon nanotubes synthesized by the chemical method have two bands—in the area of 1590 cm^{-1} (G band, tangential mode E_{2g} of symmetric valence vibrations of sp^2—hybridized carbon bonds) and about 1340 cm^{-1} (D band due to fully symmetric vibrations of A_{1g} aromatic sp^2—hybridized carbon rings). The classic spectrum of carbon nanotubes with two characteristic peaks at 1298 and 1586 cm^{-1} corresponding to D and G modes, respectively (Fig. 1.26a–c).

At the same time, on the first, second, and fourth spectra (Fig. 1.26a–c), the classical spectra for nanotubes with characteristic D and G modes are clearly observed. However, it was found that in all three samples, the D mode has a higher intensity than the second mode, which is not characteristic of intact single-walled nanotubes. That is, we can conclude that the nanotubes obtained by the CVD method are multi-walled and most likely have a curved structure. In all three samples, a shoulder in the G mode is clearly visible (which may be an overtone of the D mode) and indicates the high defectiveness of the obtained samples. The most important characteristic is the measurement of the ratio of the D/G modes (Table 1.3), which determine the defectiveness of the structures of carbon nanomaterials. According to this ratio, all nanotubes have a high number of defects and bends, but sample number 2 is the best (smallest number of defects) compared to the others.

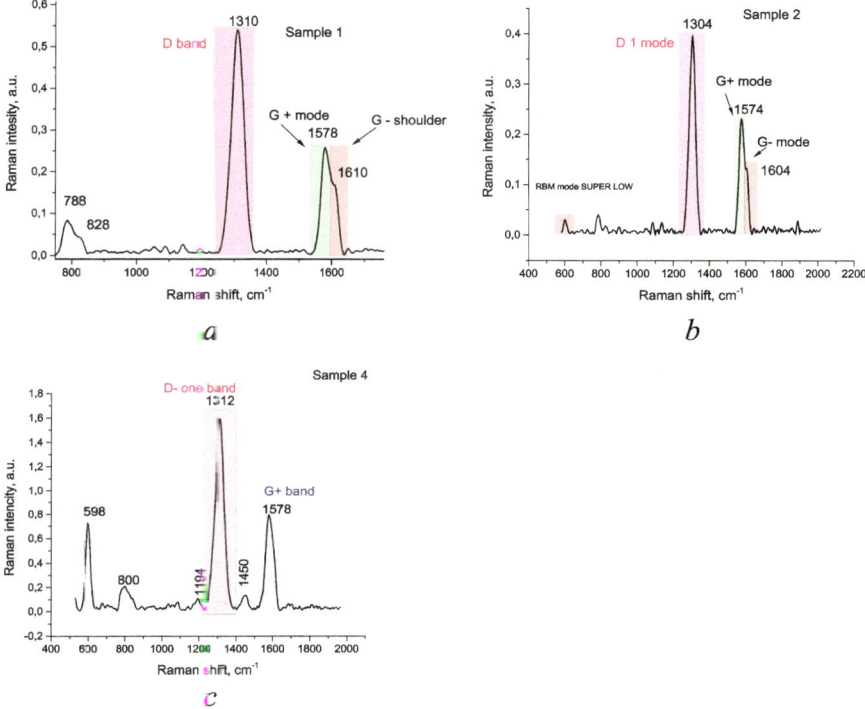

Fig. 1.26 Raman scattering spectra of carbon nanotubes obtained at different synthesis regimes: **a** sample 1, **b** sample 2, **c** sample 4. *RBM—radial breathing mode

Table 1.3 Measurement of the ratio of the D/G modes

Sample, No	$X = \frac{I_D}{I_G}$, arb. units	Method
1	2.1	CVD
2	1.7	CVD
3	–	CVD
4	2.01	CVD

According to the half-width of peak parameter D, it can be clearly seen that sample 2 has the smallest value, which indicates its highest crystallinity. Together with the conclusion made in the previous point at considering the characteristic mode intensity ratio, means that sample 2 is the best sample among those obtained.

Summary

Analysis of the averaged, kinetic, experimental data showed a rather high efficiency of the process of carbon nanomaterial growth. A carbon material was obtained in the form of multi-walled CNTs.

Raman spectra confirmed the presence of carbon nanotubes with two characteristic peaks at 1310 and 1578 cm^{-1} corresponding to D and G modes, respectively. Based on the ratio of the two modes, the ratios of the planes under the G and D bands were calculated, and a significant number of carbon nanotube defects were determined. According to the half-width peak of parameter D, it can be clearly determined that sample 2 has the smallest half-width of peak, which indicates its highest crystallinity. All nanotubes, according to the calculated ratios, have a high number of defects and bends; however, sample number 2 has the most perfect (smallest number of defects) compared to the others.

Catalyst material-iron ore pellets are affordable and inexpensive materials that do not require additional preparation. Using it, it is possible to obtain CNTs in large quantities with high efficiency and adequate cost.

1.6 Carbon Nanotubes Growth in Converted Gas Atmosphere on Dispersed Iron Catalyst Obtained as Result of Ferrocene Decomposition

Carbon nanomaterials, including carbon nanotubes, are recognized as one of the most promising materials in modern technologies due to their unique mechanical, thermal, and electrical properties. The range of applications of CNTs is constantly growing and presents tremendous opportunities to synthesize these materials. Of the many ways to obtain CNTs, one of the most potent is the CVD method [75–77]. The low cost of the process, the comparative purity of the material (without inclusions of amorphous carbon), as well as the possibility of scalability to an industrial scale make it very attractive. At the Gas Institute of the National Academy of Sciences of Ukraine, the fundamentals of the technology for producing CNTs by CVD method from the products of air conversion of hydrocarbons on catalysts based on freshly reduced iron have been developed [13, 78].

In studying these processes, much attention is paid to the preparation of substrates with a catalyst. This is one of the most time-consuming stages of the work. It is known from the literature [79–81] that ferrocene can be used as a source of dispersed iron particles. The idea of replacing traditional iron catalysts with ferrocene was tested in this study. Ferrocene ($C_{10}H_{10}Fe$) is one of the most popular representatives of metal–organic compounds. Ferrocene represents orange crystals with a melting temperature of 173 °C and a boiling temperature of 249 °C, density—1.49 g/cm^3. Decomposition of ferrocene is guaranteed to occur at the temperature of the process of CNTs synthesis—650 °C. Iron particles formed as a result of ferrocene decomposition and represent a catalyst already dispersed to a nanoscale state and do not require additional processing. This study aimed to obtain CNTs on an iron catalyst formed from the decomposition products of ferrocene in a converted gas atmosphere.

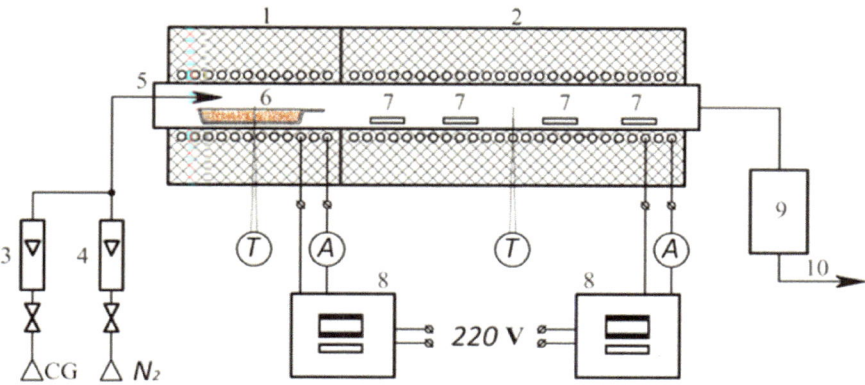

Fig. 1.27 Scheme of an experimental unit for studying CNTs deposition on dispersed iron particles in a converted gas atmosphere: *1*—electric furnace for evaporation of ferrocene; *2*—electric furnace for zone heating of CNTs growing; *3, 4*—flowmeters and valves for converted gas and nitrogen supplying; *5*—quartz reactor; *6*—boat with ferrocene; *7*—ceramic rods—deposition centers; *8*—control system; *9*—filter; *10*—outcoming gas

Scheme of an experimental unit for studying CNTs deposition. A schematic of a laboratory unit for CNTs synthesis is shown in Fig. 1.27. The installation consists of two furnaces connected in series through a gas-tight quartz reactor with electric heating. The flow rate of gas mixtures was monitored using flowmeters. In furnace 1, a boat was placed with powdered ferrocene weighing from 2 to 6 g. Ceramic rods were placed in a quartz reactor, which serves as centers for the deposition of catalyst particles. Before starting the experiment, the system was purged by nitrogen.

Furnace 1 was heated to the beginning temperature of ferrocene evaporation—270 °C, furnace 2 to the temperature of CNTs deposition—650 °C. The nitrogen flow passing through furnace 1 captured ferrocene vapors and transferred them to furnace 2. Next, furnace 2 was preheated to the temperature of the process, where ferrocene vapors were decomposed with the formation of iron particles—a catalyst for CNTs growth. In the same place, the precipitation of dispersed catalyst particles on the deposition centers was carried out, followed by removing gaseous products of ferrocene decomposition using nitrogen as a carrier gas. Thereafter, the nitrogen was replaced by converted gas. Converted gas was fed through a piping system from a converted gas generator (Fig. 1.28).

The conversion of natural gas was carried out in a converter on a nickel catalyst. The converter is heated by an electric furnace; the temperature in the catalyst layer was about 950 °C. The conversion of natural gas was carried out with air heated to a temperature of 550 °C in an air heater 2, and the air flow rate was about 0.3. The resulting converted gas at first was cooled in a cooler 3 and then dried in a moisture absorber 5 filled with a silica gel. The dried gas was fed into CNTs synthesis unit, in a quartz reactor in which iron particles from ferrocene were deposited on ceramic rods. Earlier it was found that the highest rate of carbon yield occurs at a temperature of about 650 °C; therefore, at the experiments, the reactor was preheated

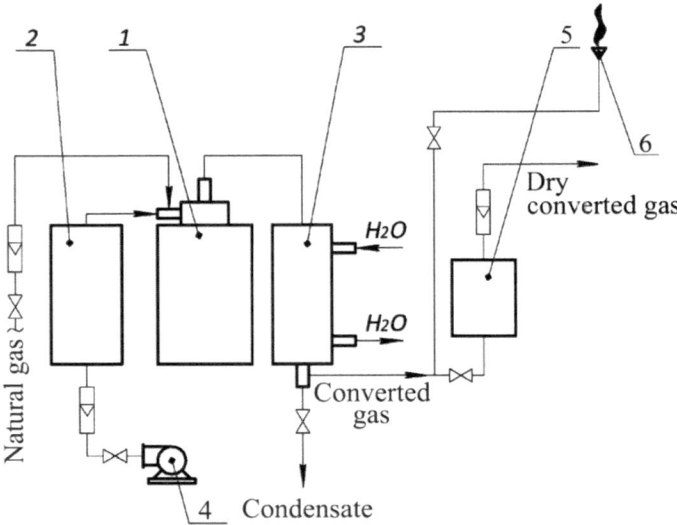

Fig. 1.28 Scheme of an experimental air converted gas generator: *1*—converter; *2*—air heater; *3*—cooler; *4*—air compressor; *5*—moisture absorber; *6*—bleeder valve

to this temperature. The converted gas discharge was about 6 l/h. In the course of the experiment, gas samples were taken for chemical analysis before and after the CNTs reactor, which were analyzed on an Agilent 6890N gas chromatograph, as well as on a Gazochrome 3101 instrument.

Carbon is deposited in the form of CNTs on an iron catalyst at a temperature of about 650 °C from the converted gas. The last in the general case is a mixture of hydrogen, carbon monoxide, and nitrogen. The experiments were carried out for two to four hours. After the end of the process, the system was cooled by nitrogen in order to prevent the deposition of foreign carbon materials from the CG atmosphere when the temperature was lowered. The resulting material was collected from the deposition centers and the walls of the reactor, weighed, and analyzed using an electron microscope (SEM Carl Zeiss Sigma 300).

Temperature dependence of carbon material yield on reduced iron. Ferrocene decomposition occurs at a temperature above 500 °C. In this case, the process by itself is described by the reaction (1.31):

$$Fe(C_5H_5)_2 \rightarrow Fe + H_2 + CH_4 + C_5H_6 + \cdots \qquad (1.31)$$

In our case, ferrocene vapor, getting into the hot zone of the reactor, decomposes with the formation of dispersed iron particles, which are deposited on ceramic rods—the crystallization centers.

Iron is the only reaction product in a solid aggregation state, while gaseous decomposition products are carried away by a flow of nitrogen. Iron particles settle on ceramic rods and, in turn, become centers of CNTs growth. The thermodynamic conditions of the reaction are responsible for the deposition of carbon on iron particles in the form of CNTs. While heating the atmosphere containing CO, the formation of solid carbon occurs due to the process of disproportionation of carbon monoxide according to the reaction:

$$2CO = CO_2 + C. \tag{1.32}$$

Reduced iron, nickel, cobalt, and palladium are the catalysts for this reaction. Research has shown that fresh-reduced iron produces the best results. The temperature of the process was determined according to our earlier studies, at which the process of carbon formation proceeds at the maximum rate. This temperature (650 °C) was taken as the "base" one.

The dependence of carbon content in the iron catalyst on the temperature of the process is shown in Fig. 1.17, Sect. 1.4. From the diagram, it follows that at the experimental conditions, the maximum rate of carbon yield was observed at a temperature of about 650 °C. At lower temperatures, the process of carbon formation is inhibited by kinetic reasons due to the sharp drop in the rate of chemical reactions. As the temperature rises, the rate of chemical reaction increases sharply, but at $t > 650$ °C, a carbon monoxide deficiency may occur, which may participate in carbon formation reactions.

Usually converted natural gas contains free hydrogen, the content of which is almost twice the content of CO and reaches 35%.

An analysis of thermodynamic calculations led to the conclusion that the formation of carbon in the presence of hydrogen occurs due to an exothermic reaction:

$$CO + H_2 = C + H_2O \tag{1.33}$$

Taking into account that the formation of the same amount of carbon by reaction (1.33) requires two times less carbon monoxide than by reaction (1.32), the initial addition of hydrogen leads to a significant increase in the thermodynamic possibility of carbon formation. Thus, the converted gas used in this work, which is a mixture of hydrogen and carbon monoxide, is an effective source of CNTs growth.

Initially, the resulting material was deposited on alumina wadding. Unfortunately, the high contamination of the obtained carbon material with alumina wadding made it practically impossible to separate the target product. Therefore, it was decided to use ceramic rods as substrates, on which iron particles were supposed to settle, and, consequently, carbon in the form of CNTs.

In our previous works, we obtained the catalyst by the method of reduction of iron (II) oxide at low temperatures. The effect of the catalyst reduction temperature on the size of the formed iron particles was observed. With a decrease in the reduction temperature, the size of the formed iron grains decreases, while the mobility (self-diffusion) of its atoms (ions) and, as a consequence, the process of collective recrystallization and enlargement of particles sharply slows down. The resulting material acquires a developed surface and consists of small differently oriented crystals of iron, the lattice of which contains many defects. Due to the uncompensated chemical bonds, iron atoms have increased activity, which is manifested as pyrophoricity. Boundaries are formed between the contacting grains of the metal according to the type of "Mott island model". According to this model, at the grain contact boundary, the order in the arrangement of atoms can be largely absent—such boundaries are called incoherent. It is known that the greatest number of defects in the crystal lattice of iron occurs at incoherent boundaries of contact of grains in contact with each other. It is these areas that have increased catalytic activity, and on them, first of all, the release of carbon atoms occurs during the decomposition of CO with the subsequent formation of nuclei of a new solid phase. The nucleation of a new phase (nucleation process) along grain boundaries is also facilitated by a higher diffusion rate of carbon atoms in these places. The iron atoms located on the surface of the crystal lattice, in contrast to those inside, create tension due to the uncompensated bonds between them. In the places of contact of the contacting iron grains, the effect of this tension is enhanced, which, apparently, also contributes to an increase in the catalytic activity of these places and the appearance of a new phase on them.

However, preparation of a catalyst based on freshly reduced iron, that is, its preliminary oxidation followed by reduction to obtain particles with a developed surface and, accordingly, high chemical activity, is a laborious and energy-consuming process.

Further work showed the validity of this assumption: the walls of the quartz reactor were practically clean, and most of the material was deposited on ceramic rods. Figure 1.29 shows SEM images of the obtained material. Unfortunately, due to objective reasons, the photos turned out to be not very presentable. Nevertheless, a comparative visual analysis with the results of other authors, for example [82], shows the similarity of the obtained carbon material, namely carbon nanotubes with a diameter ~ 100–200 nm. The authors plan to continue their R&D related to the CVD method for the production of carbon nanotubes and nanofibers on various metal catalysts.

When using ferrocene weighing 3 g, the yield of carbon material forms 13 g. That is, taking into account the fact that iron in the ferrocene molecule is 30% by weight, it turns out that on one gram of iron formed as a result of ferrocene decomposition, carbon is deposited in the form of CNTs weighing 14 g. The ratio of the mass of the settled carbon and the mass of the iron catalyst indicates the high efficiency of the applied method.

Summary

In the course of the study, an attempt was made to evaluate and test on a laboratory scale the technology for producing CNTs by the CVD method using ferrocene

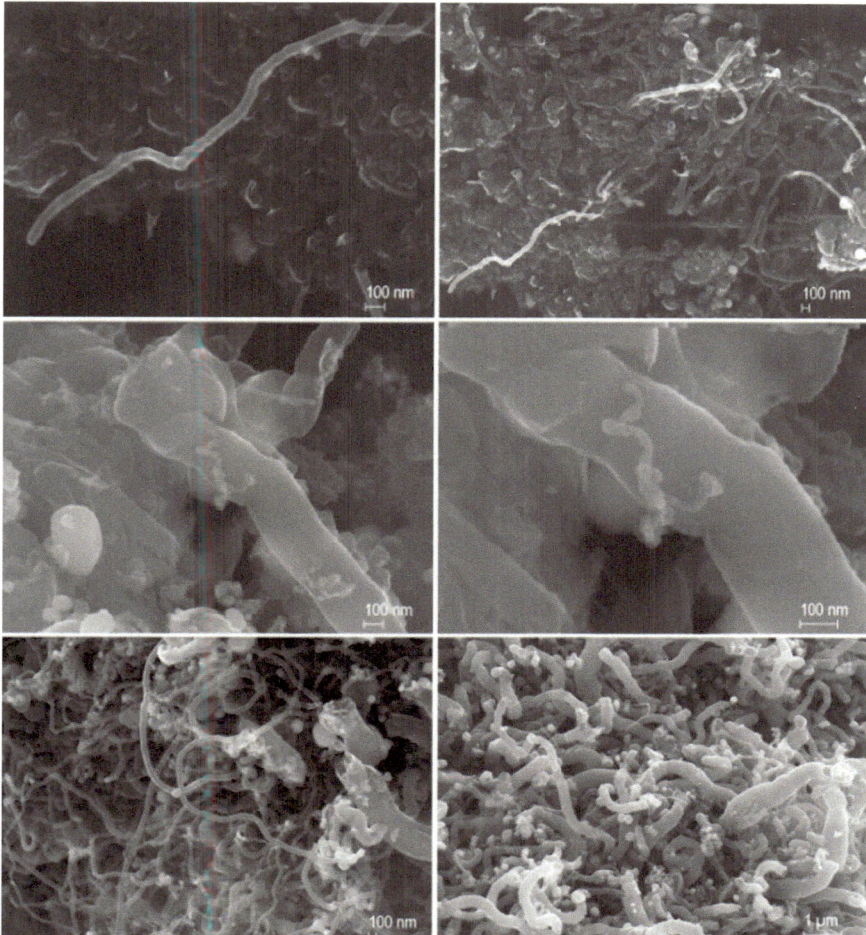

Fig. 1.29 SEM photo of obtained CNTs

decomposition products as a catalyst. The products of air conversion of natural gas served for CNTs synthesis. Target carbon was obtained by disproportionation of carbon monoxide. The temperature regime for CNTs growth was about 650 °C and was chosen as optimal for the iron catalyst. Only precise control of the composition of the gas mixture and the ratio of the carbon source—carbon monoxide to hydrogen, as well as the temperature regime allows CNTs obtaining practically without external carbon impurities in the form of amorphous carbon.

Experiments have shown that dispersed iron particles obtained from ferrocene have a high reactivity, in many respects not inferior to freshly reduced iron, which, due to its low reduction temperature, has a developed surface and, as a consequence, is one of the most effective catalysts for carbon deposition. But the process of preparing such a catalyst is a rather laborious and energy-consuming process, which consists in multiple repetitions of oxidation–reduction cycles in order to obtain the most developed substrate surface for CNT deposition.

As a percentage of one unit of iron, approximately 14 weight units of carbon were obtained in the form of CNTs. The results on the yield of the target carbon product confirmed the possibility of using ferrocene as an effective catalyst for the growth of CNTs, as well as the possibility of using this technology on an industrial scale.

1.7   Formation of Carbon Nanotubes from Products of Methane Air Conversion on Ni/Cr and Fe Catalysts

In this part of work is to determine the mechanisms of formation of carbon nanotubes on substrates with Fe and Ni/Cr nanostructures. The promoters of nanotube growth were deposited by thermal evaporation in a vacuum universal post as thin polydisperse layers of metals. The surface layers of iron and chromel were deposited on sitall (glass–ceramic) substrates. X-ray phase analysis of as-deposited catalysts showed the formation of growth centers of nanotubes. Carbon nanomaterials with different morphology were synthesized from methane air conversion products by chemical vapor deposition. The Raman spectra showed the defective nature of the deposited materials.

Today, nanocarbon materials of different types attract the attention of researchers and developers of new devices in various fields of science and technology due to a unique combination of electrical, mechanical, thermal, optical, and other properties. In particular, they are considered as the materials for building blocks for the development of new nanostructured materials for the functional elements of electronic and photonic devices. These materials improve mechanical, thermal, and interfacial properties of multiphase polymeric composites graphene/carbon nanotubes (CNTs) [83].

The optimal functional characteristics of nanocarbon materials are directly related to their internal structure which can vary significantly depending on the conditions of their production and additional treatment. Therefore, the information about their structure and its dependence on different methods of impact is important for targeted development of materials with specified physical and mechanical properties. Data on the effect of the structure of nanocarbon materials on their optical characteristics is key in determining the conditions for producing samples with improved functional parameters for photonic elements.

Acetylene, methane, ethylene, propylene, and benzene are most often used for the synthesis of CNTs from hydrocarbons. It is known that effective catalysts for the formation of CNTs are metals of the iron subgroup—Fe, Co, Ni, and their mixtures. SiO_2 and Al_2O_3 are usually used as catalyst carriers. As a carbon source, we chose a conversion gas with a significant content of carbon monoxide (CO). One carbon atom per molecule provides significant control over the technological parameters of carbon materials, namely the precursor concentration, flow rate, partial pressure, temperature, the type of catalyst, and so on. Researchers have described in detail the $Co–Mo/SiO_2$ catalysts that were used for the synthesis of single-walled CNTs using the CO disproportionation reaction at 700–850 °C. It is believed [84] that the use of carbon monoxide made it possible to produce single-layer nanotubes. The thermal decomposition of pure CO on a Ni–CO catalyst deposited on Al_2O_3 was performed at a pressure slightly above atmospheric pressure and at 1200 °C. Single-layer CNTs were also produced.

A great achievement was the synthesis of molybdenum catalyst used for the synthesis of CNTs from CO. The catalyst was produced by impregnation of alumina with a solution of molybdenum derivatives in methanol, followed by heating to 200 °C. The diameter of CNTs formed on this catalyst was 1–5 nm. At the ends of many of them, catalyst particles several nanometers in size were found, which allowed the researches to suggest the "yarmulke" mechanism for coating growth. The film of chemisorbed carbon atoms on the surface of a catalyst particle is called a "yarmulke". According to this mechanism, the catalyst particles promote the dehydrogenation of hydrocarbon molecules deposited on them from the gas phase. Carbon diffuses to the open end of a CNT (where the catalyst particle is located) along the surface or in the volume of the catalyst and fits into the CNT structure. The systems Ni–MgO, CO–MgO, and $Ni–AlPO_4$ were also successfully tested as catalysts for the decomposition of CO [85].

This paper presents the results of producing nanocarbon materials. This topic belongs to the priority fields of scientific research, development, and technology in Ukraine.

Scheme of technological unit for producing thin nanocomposite films by vacuum co-deposition. Iron [86] and standard chromel K alloy (89–91 Ni, 8.5–10 wt.% Cr) [87] catalysts were deposited on sitall substrates [88] in a modernized vacuum universal post VUP-5M (Fig. 1.30) by vacuum thermal evaporation of iron and chromel at a residual gas pressure of $10^{-2}–10^{-3}$ Pa, field strength $E = 60–80$ V/cm^2, and substrate temperature 300 °C. The layers were formed on the surfaces of polished sitall substrates pre-cleaned by a standard chemical method [89]. The substrates were positioned in a separate zone of the vacuum post. The samples were located at a distance of $h = 8 \times 10^{-2}$ m above the evaporator, which promoted a uniform deposition of iron and chromel. The advantage of this technique is the ability to use two evaporators in one cycle and to apply annealing without the need to depressurize the chamber. A system of automatic control of the film growth was developed to control the reproducibility of a given distribution of components through the thickness of the films [90].

The control of evaporated material was performed by monitoring the shift of the resonant frequency of quartz sensors (8 MHz), the parameters of electric arc evaporation, and the substrate temperature. The signals of the quartz resonant sensors that registered the mass of each film component were used to control the molecular fluxes in real time. The sensitivity of quartz sensors was 6×10^{-8} kg/m^2, and the software-controlled frequency drift did not exceed 10 Hz/h. The synthesis of carbon

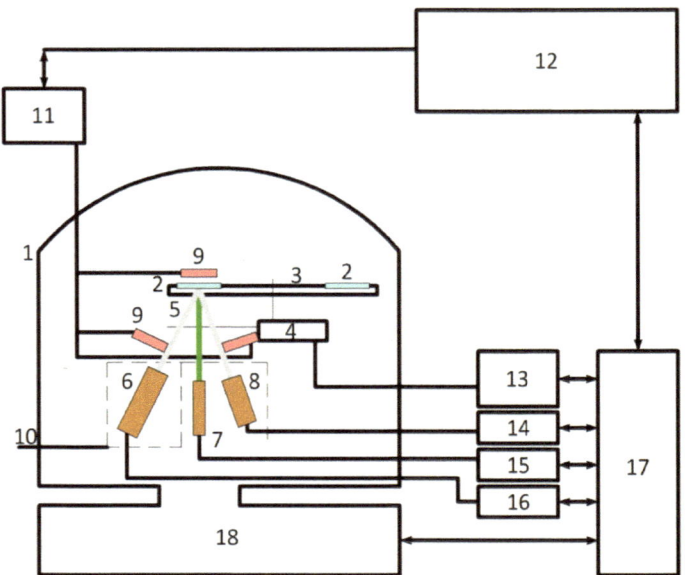

Fig. 1.30 Scheme of technological unit for producing thin nanocomposite films by vacuum co-deposition: *1*—vacuum chamber; *2*—samples; *3*—substrate transport system; *4*—transportation system engine; *5*—shutter; *6, 7, 8*—metal evaporators; *9*—quartz sensors for controlling molecular beams and film mass; *10*—light guides; *11*—quartz resonator controller; *12*—ADC, DAC, coupling board; *13*—pressure sensor with controller; *14, 15, 16*—power supply; *17*—converters and manipulators; *18*—vacuum unit VUP-5M [90]

nanotubes was carried out using as raw material one of the most affordable and cheap hydrocarbon gas–natural gas. The main component of natural gas is methane; its content can reach 90% or more.

The bond in the CH_4 molecule between carbon and hydrogen atoms is relatively strong, so the decomposition of methane, even with a catalyst, requires elevated temperatures which provide the preconditions for the formation of soot carbon together with the nanotubes. In order to convert the carbon contained in CH_4 molecules into a less stable compound, natural gas is pre-converted by transforming methane into unstable carbon monoxide (CO) at low temperatures. The use of natural gas conversion products made it possible to reduce both the synthesis temperature of carbon nanotubes to 600–650 °C and the content of amorphous carbon in these nanotubes.

The conversion of natural gas was performed at the converter developed and constructed at the Gas Institute of the National Academy of Sciences of Ukraine (Fig. 1.21, Sect. 1.5.1).

The converter *1* was heated by an electric furnace, and the temperature in the catalyst film was about 950 °C. The conversion of natural gas was carried out with air heated in the air heater *2* to 550 °C; the air flow rate was about 0.3. The converted gas was first cooled in the cooler *3*, and then dried in the desiccant *4* filled with silica

Fig. 1.31 Scheme of experimental unit for producing carbon nanomaterials

gel. As a result, the converted gas contained up to 18% CO and 35% H_2, 1.5–2% oxidants, and 0.5–1.5% CH_4, N_2 balance. The gas composition was measured using a Gazochrom gas analyzer.

The dried converted gas was fed into the unit for CNT production, where the carbon material was grown on sitall substrates with deposited catalyst layer in the quartz reactor. It was determined in [91] that the highest rate of carbon yield occurs at about 650 °C, so the reactor was preheated to this temperature in our experiments. The consumption of converted gas was about 6 l/h.

The carbon nanotubes were grown in a horizontal quartz reactor with external electric heating (Fig. 1.31).

As the catalyst carriers, sitall plates were used; nanosized films of iron, chromel, and nichrome were deposited on the surfaces of these plates. The plates were sequentially placed in the center of the reactor. The substrate coated with Fe was oxidized at $T = 600$ °C in air atmosphere for 30 min. After oxidation, the reactor was blown down with nitrogen. Oxidized plates were reduced in a hydrogen atmosphere for one hour with a gradual temperature increase from 400 to 650 °C. Reductive treatment of iron oxides (on iron plates) with hydrogen at relatively low temperatures enables to get a surface of freshly reduced iron with high catalytic activity. After hydrogen treatment, the gas atmosphere in the reactor was replaced by the products of natural gas air conversion. The converted gas was continuously fed into the reactor heated up to 650 °C for 2.5 h.

After the deposition of the carbon nanomaterial, the reactor with the plates in it was blown down with nitrogen and cooled to 80 °C. The sequence of heat treatment of the material in different gas atmosphere is shown in Fig. 1.32.

X-ray phase analysis was carried out with an X-ray diffractometer Ultima-IV, Rigaku, Japan (National Technical University of Ukraine "Igor Sikorsky Kyiv Polytechnic Institute") in CuK_α radiation. The peaks in diffraction patterns were identified using the databases ICDD PDF-2 and PDF-4. The carbon nanomaterials were investigated with scanning electron microscopy (JSM-6490LV microscope) with a resolution of up to 1 nm (Fig. 1.33). The Raman spectra of the samples were investigated using a Raman spectrometer mini-Raman Pro (Lightnovo, Denmark). A laser with wavelength $\lambda = 785$ nm was used for the spectrum range 600–2000 cm^{-1}. The

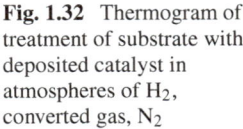
Fig. 1.32 Thermogram of treatment of substrate with deposited catalyst in atmospheres of H_2, converted gas, N_2

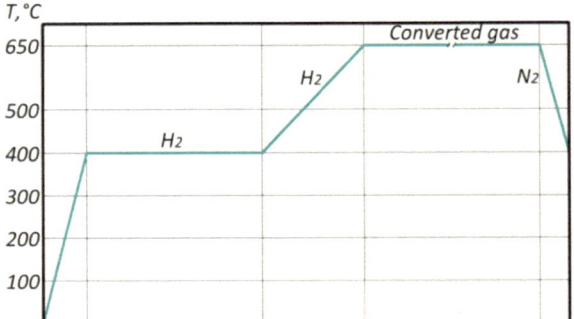

accuracy of determining the wave number was in the range 0.5–2 cm^{-1}. The registration parameters were the following: exposure 100 ms, 50 spectra were averaged. The laser power was 5.88–47.38 mW.

The catalyst layers of chromel and iron were deposited under the conditions described above. The deposition process was provided by the possibility to deposit several materials without depressurizing the chamber, with little thermal impact on the treated structure. To control the reproducibility of films with a given distribution of components through the thickness, a system of automatic control of the process of film growth was developed.

The synthesis and growth of the films were controlled with two computerized control systems. These measures provided the production of films with reproducible technical parameters. The mass scatter of the samples in the experiments of the same type of did not exceed 15%. Diffraction maxima of nickel and chromium were observed in diffractograms from chromel film (Fig. 1.34). Apparently, during the deposition of chromel, it decomposed with the formation of metals and chromium oxide (Cr_2O_5). X-ray spectrum analysis also revealed the signals that correspond to the structure of nickel chromate $NiCr_2O_4$. According to [92], nickel chromate can be the center of growth of carbon nanotubes and is a promising catalytic material along with pure metals. Oxides of silicon and aluminum are considered as the contribution from the substrate. The most intense diffraction maxima of chromium oxide Cr_2O_5 are observed in the ranges $2\theta = 10$–$18°$ and 28–$30°$.

From the viewpoint of thermodynamics, the reaction of thermal decomposition of CO (with CO_2 and carbon as products) crucially differs from the processes of pyrolysis of hydrocarbons: the equilibrium yield of carbon at atmospheric pressure approaches noticeable values in the region of low temperatures 300–750 K, and it reduces as temperature increases and pressure decreases. On the contrary, the carbon yield in the pyrolysis of C_2H_2 and CH_4 increases with increasing temperature and decreasing pressure, approaching noticeable values at 1250–1500 K. Therefore, it would seem that carbon monoxide is a less convenient starting reagent for the synthesis of CNTs. However, it is more difficult to heat hydrocarbons above 800–900 °C before contact with the catalyst as compared to CO for kinetic reasons. In this regard, CO has an advantage over hydrocarbons.

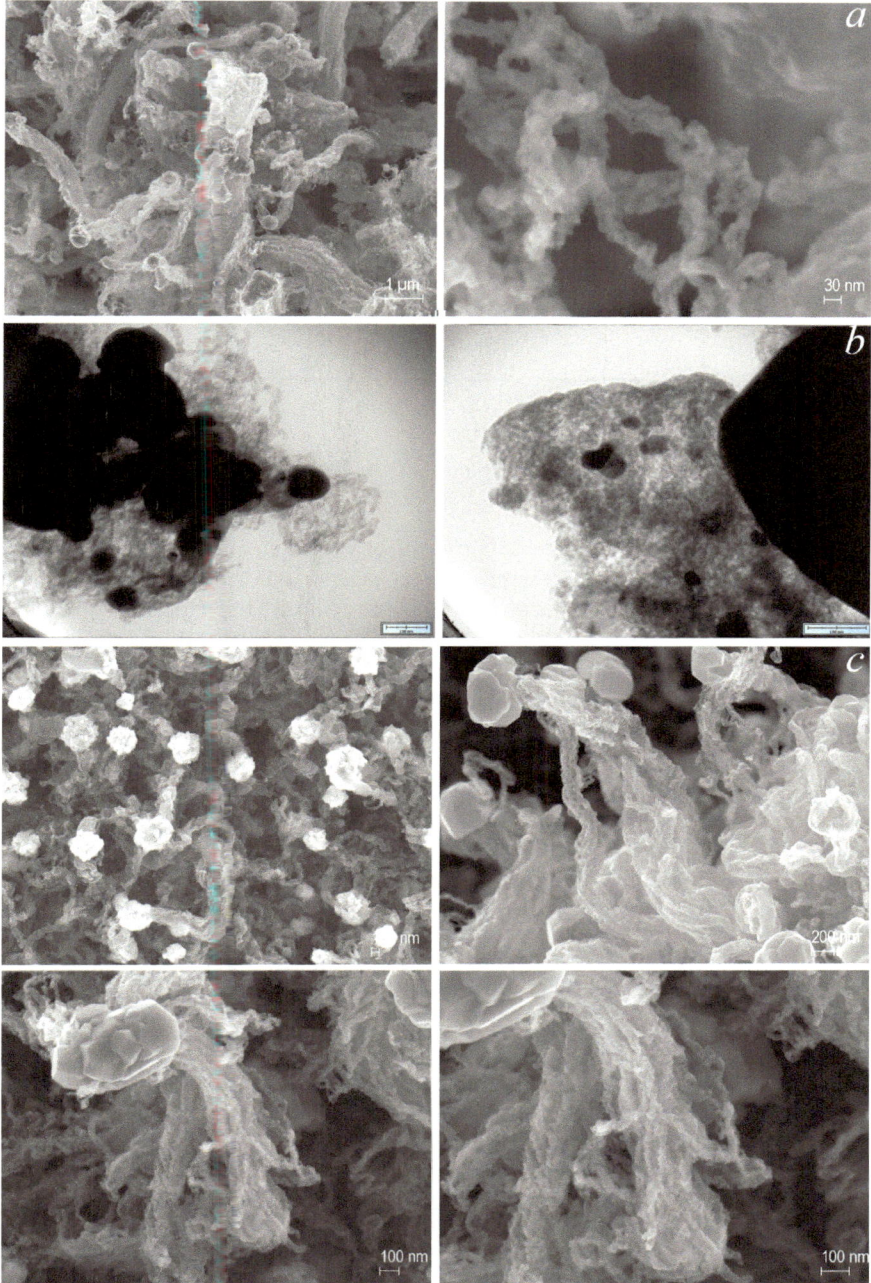

Fig. 1.33 Micrographs of carbon nanomaterials: **a, b** iron catalyst (SEM and TEM); **c** chromel catalyst (SEM)

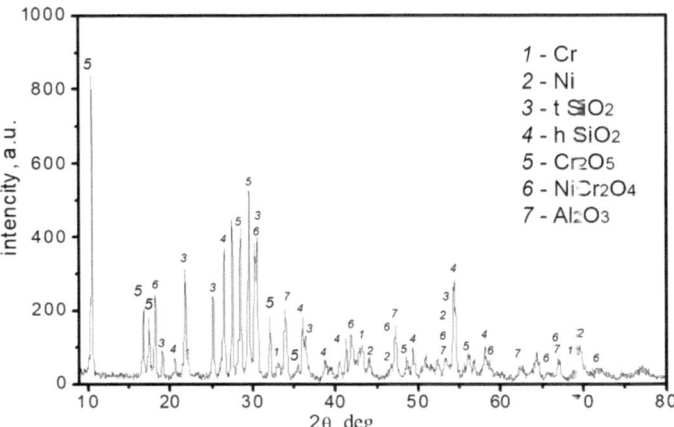

Fig. 1.34 Diffractogram of thin chromel layer on sitall substrate

Carbon formed on metals of the subgroup of iron and their alloys with other metals appears in the form of deposits with different morphological and crystallographic characteristics that depend on the properties of the catalysts and experimental conditions. These dependencies make it possible to purposefully control the technology of obtaining carbon deposits with different morphology. Carbon can be deposited as threads of various configurations, nanotubes, plates, and so on [93]. The mechanisms of catalytic formation of CNTs are divided into root growth and tip growth [94]. In the first, CNTs grow from the surface of the metal particles that remain in contact with the substrate. In the second, a metal particle detaches from the substrate and is held on a growing nanotube. The synthesis of CNTs from arc discharge or by laser dispersion in most cases occurs by root growth [95].

1.7.1 Raman Spectra of Carbon Materials with Iron Catalyst, Chromel Catalyst, Multilayer Nanotubes

Raman spectroscopy is a technique for rapid and non-destructive monitoring of electron and phonon parameters of carbon materials, which correlate with their chemical composition and structure at the level of chemical interactions. The Raman spectra (Fig. 1.35) of the produced materials are characterized by the position, half-width, and relative intensity of the spectral bands, which are determined by the structural state of carbon. Typically, the Raman spectra of chemically synthesized multilayer carbon nanotubes have two bands, in the region of 1590 cm^{-1} (G band, tangential mode E_{2g} of symmetric valent vibrations of sp^2 hybridized carbon bonds), and about 1340 cm^{-1} (D band caused by totally symmetric vibrations of A_{1g} aromatic sp^2 hybridized carbon rings).The classical spectrum of carbon nanotubes contains two characteristic peaks at 1298 and 1586 cm^{-1}, which correspond to the D and G mode, respectively (Fig. 1.35a). The first mode D characterizes the defect structure of graphene layers of nanotubes. The second mode G corresponds to the tangential valent carbon–carbon vibrations and characterizes the ordering of the carbon phase. The ratio of these two modes shows the amount of defects in nanotubes. Known lines in the Raman spectra of insufficiently ordered graphite are within 1585–1570 cm^{-1} and $1350 \div 1300$ cm^{-1} [96].

The Raman spectra in Fig. 1.35b, c are almost identical. These spectra show strong luminescence in the range 1100–2000 cm^{-1}, which is not typical for the samples of purified nanotubes (Fig. 1.35c). Additional signals are observed, probably from different types of carbon and/or catalysts. This luminescence can be caused by the transparency of the sample and the contribution from the substrate (the beam hit the area where the number of nanotubes was very small); however, this hypothesis requires further study. Characteristic D and G modes at frequencies 1364 and 1510 cm^{-1} are found out. These modes can be caused by multilayer nanotubes which were observed in electron microscopy images. It was found that the D line is more

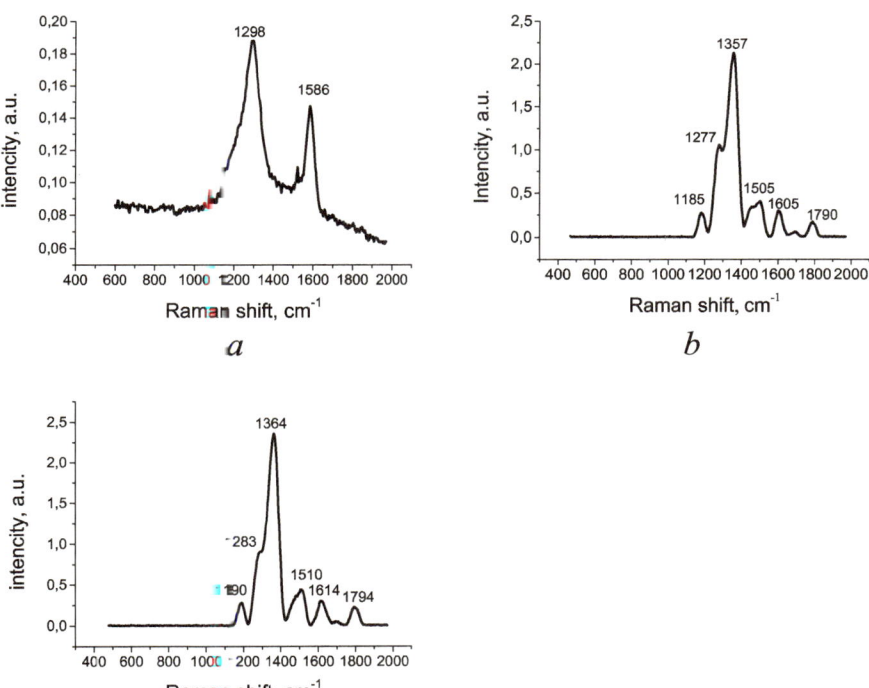

Fig. 1.35 Raman spectra of carbon materials with: **a** iron catalyst; **b** chromel catalyst; **c** multilayer nanotubes

intense and wider than the G line (Fig. 1.35), which indicates a large number of defects ($D/G < 1$). This ratio also characterizes the curvature (bending) of the layer. Besides, the G mode disintegrates into $G+$ and $G-$ (peak splitting), which indicates the presence of single-layer CNTs.

Summary

Thin layers of iron and chromel catalysts were produced in the developed unit. CNTs were synthesized from the products of methane air conversion by chemical vapor deposition in a flow reactor. It is shown that the synthesis of CNTs at $T >$ 650 °C increases the mobility of catalytic particles in the carrier, which leads to their agglomeration, and therefore affects the growth of CNTs.

The electron microscopy (SEM and TEM) studies of the structure of carbon nanomaterials showed that the developed synthesis technology makes it possible to form CNTs of different morphology and thickness. The Raman spectra confirmed the presence of carbon nanotubes with two characteristic peaks at 1364 and 1510 cm^{-1}, which correspond to D and G modes, respectively. The ratio of these two modes indicated a significant number of defects in carbon nanotubes.

1.8 Features of Formation of Carbon Nanotubes from Products of Methane Air Conversion by Chemical Deposition Method

The work is to elucidate some regularities of carbon nanotube (CNT) formation from methane air conversion products. Carbon nanomaterials of different morphology are synthesized by the method of chemical vapor deposition on metal catalysts of the iron group. Thermodynamic calculations and empirical studies allow choosing the optimal process temperature, at which the mold product contains almost no harmful impurities, namely soot. The idea of sequential treatment of a metal catalyst in oxidizing and reducing atmospheres is practiced. Raman spectra show the defective nature of the obtained materials, i.e., CNT.

Currently, nanocarbon materials of various nature, due to a unique combination of electrical, mechanical, thermal, optical and other properties, attract the close attention of researchers and developers of new devices in various fields of science and technology. In particular, they are considered as starting "building blocks" for the development of new nanostructured materials for functional elements of electronic and photonic devices. These materials provide enhanced mechanical, thermal, and interphase properties of the graphene/carbon nanotube (CNT) multiphase polymer composites [97].

Optimal functional characteristics of nanocarbon materials are directly related to their internal structure which can vary significantly depending on the conditions of their production and further processing. Therefore, information about the structure of these materials and its dependence on various processing techniques is important for the targeted development of materials with given physical and mechanical properties. Data on the effect of the structure of nanocarbon materials on their optical characteristics is key to determining the conditions for producing samples with enhanced functional parameters for photonic elements.

Most often, acetylene, methane, ethylene, propylene, benzene, etc. are used for the synthesis of CNTs from hydrocarbons. As a carbon source, we chose the products of air conversion of natural gas with a sufficient content of carbon monoxide (CO). One carbon atom per one substance molecule provides significant and important control over the technological parameters of producing carbon materials: namely over the precursor concentration, flow rate, partial pressure, temperature, catalyst type, etc.

This work presents the results of producing nanocarbon materials that correspond to the priority directions of the development of science, technology, and engineering in Ukraine.

The synthesis of carbon nanotubes. The process of synthesis of carbon nanotubes was carried out using as a raw material one of the most accessible and cheap hydrocarbon gas—natural gas. The main component of natural gas is methane; its content can reach 90% or more. The bonds between carbon and hydrogen atoms in the CH_4 molecule are comparatively strong. Therefore, elevated temperatures are required for the decomposition of methane, even with a catalyst, which provides the prerequisites for the formation of sooty carbon along with nanotubes. In order to convert carbon in the CH_4 molecule into a state with less strong bond, natural gas is preliminarily converted: methane is transformed into carbon monoxide CO which is unstable at low temperatures. The use of natural gas conversion products made it possible to reduce both the temperature of the synthesis of carbon nanotubes to 873–923 K and the content of amorphous carbon in them.

A converter constructed at the Gas Institute of National Academy of Sciences of Ukraine was used for the conversion of natural gas (Fig. 1.21, Sect. 1.5.1).

The converter was heated by an electric furnace; the temperature in the catalyst layer is about 1223 K. Natural gas conversion was carried out in air heated up to 823 K in the air heater 2; the air flow rate was 0.27–0.3. The converted gas (CG) was first cooled in the refrigerator 3, and then dried in the moisture absorber 5 filled with silica gel. The converted gas contained up to 19% CO and 35% H_2, 1.5–2% oxidizers, 0.5–1.5% CH_4, N_2 balance. The gas composition was measured using a Gasochrom gas analyzer.

The dried converted gas was fed to the CNT production unit (Fig. 1.36) where the carbon material was grown in the quartz reactor on steel plates. According to [98], the highest rate of carbon release occurs at about 923 K; therefore, the reactor was preheated to this temperature. The consumption of converted gas was about 6 l/h.

The carbon nanotubes were grown in a horizontal quartz reactor with external electric heating (Fig. 1.36).

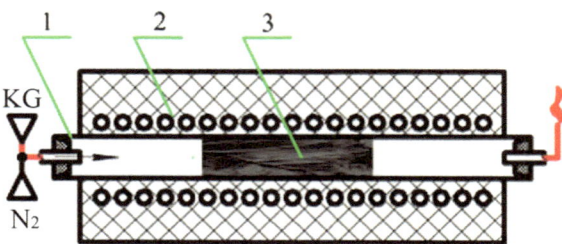

Fig. 1.36 Experimental unit for deposition of carbon nanomaterials (scheme): *1*—gasproof quartz reactor with spent gas afterburning; *2*—electric split furnace with automated control; *3*—steel catalyst plate for CNT deposition

The CNT deposition lasted for two hours. After the deposition, the reducing atmosphere of CG was replaced by an inert atmosphere (nitrogen), and the samples were cooled to room temperature. Cooled samples were cleaned of CNTs. The synthesized carbon nanomaterials were studied by means of scanning electron microscopy (SEM) at a JSM-6490LV microscope.

The Raman spectra (RS) of the samples were studied using a mini-Raman Pro spectrometer (Lightnovo, Denmark). A laser with a wavelength of $\lambda = 785$ nm was used with a spectral range of 600–2000 cm^{-1}. The accuracy of determining the wave number was 0.5–2 cm^{-1}. The exposure time was 100 ms; 50 spectra were averaged. The power of the lasers was 5.88–47.38 mW.

1.8.1 Structural Features of Carbon Nanomaterials Synthesized on Steel Plates Depending on Their Previous Processing

Pre-treated steel plates were used as catalysts. The metal plate at different stages of the technological process is shown in Fig. 1.37.

The plates were oxidized with air oxygen; their reduction was carried out in an atmosphere of hydrogen or converted gas. This process was carried out in order to study the effect of preliminary activation of the surface of the plates on the deposition

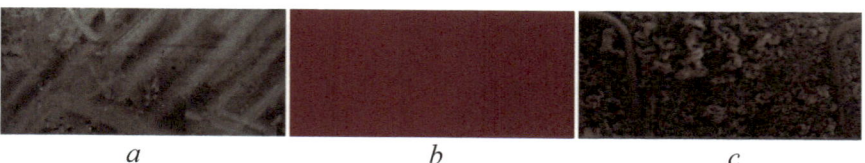

a b c

Fig. 1.37 Steel plate used as a catalytic substrate: **a**—initial state; **b**—oxidized plate; **c**—plate with deposited carbon nanomaterial

of CNTs and their morphology. Five test plates were produced. Figure 1.38 shows the morphology of the deposited carbon nanomaterials. Plates (a) and (d) (Fig. 1.38) were oxidized at $T = 823$ and 923 K, respectively, for thirty minutes in an air atmosphere and reduced with converted gas. After the surface was cleaned from carbon material, the plates were again oxidized in air at the same temperatures. After reoxidation, the samples were reduced in a hydrogen atmosphere at $T = 673$ K for 20 min. Plates (b) and (c) were oxidized in one step for thirty minutes at 923 and 973 K, respectively. Oxidized plates and the plates without pretreatment were placed in the quartz reactor *1* of the experimental unit (Fig. 1.21, Sect. 1.5). Converted gas was fed into the reactor; the CNTs were synthesized in its atmosphere at 923 K for two hours.

According to [99], the thermodynamics of the reaction of CO thermal decomposition (disproportionation into CO_2 and carbon) drastically differs from the pyrolysis of hydrocarbons. The equilibrium yield of carbon at atmospheric pressure reaches considerable values at low temperatures (300–750 K), and it reduces with increasing temperature and lowering pressure. On the contrary, carbon yield during pyrolysis of C_2H_2 and CH_4 increases with increasing temperature and decreasing pressure; it reaches considerable values at 1250–1500 K.

In this regard, carbon monoxide would seem a less convenient starting reagent for the synthesis of CNTs. However, it is more difficult to heat hydrocarbons above 1073–1173 K as compared to CO before contact with the catalyst due to kinetic reasons, so the latter has advantages.

Carbon deposited on the metals of the iron subgroup and their alloys has different morphological and crystallographic characteristics which depend on the type of the catalysts and the conditions of the deposition (Fig. 1.38a–d). Therefore, it is possible to purposefully control the technology of forming carbon structures with different morphologies. Carbon can be deposited in the form of fibers with different morphology, nanotubes, plates, "nanohorns", etc. [93].

The mechanisms of catalytic growth of CNTs are distinguished as "root" and "tip" ones [100]. Upon the root growth, CNTs grow from the surface of a metal particle that preserves contact with the substrate. Upon the tip growth, a metal particle detaches from the substrate and is located on the growing nanotube. The root growth occurs during the synthesis of CNTs from arc discharge and during laser deposition [95].

Raman spectroscopy (RS) is a technique for fast and non-destructive control of electronic and phonon parameters of carbon materials, which correlate with their chemical composition and structure at the level of chemical interactions. In the Raman spectra (Fig. 1.39) of the deposited materials, the position, half-width, and relative intensity of the spectral bands are of interest; these parameters are determined by the structure of carbon. As a rule, the Raman spectra of multilayered carbon nanotubes synthesized by the chemical technique have two bands: at about 1590 cm^{-1} (G band, tangential mode E_{2g} of symmetric valence vibrations of sp^2 hybridized bonds of carbon), and at about 1340 cm^{-1} (D band associated with totally symmetric vibrations A_{1g} of aromatic sp^2 hybridized carbon rings). The typical spectrum of carbon nanotubes contains two characteristic peaks at 1298 and 1586 cm^{-1}, which correspond to the D and G modes, respectively (Fig. 1.39). The first D mode characterizes the defectivity of graphene layers of nanotubes. The second G mode corresponds to

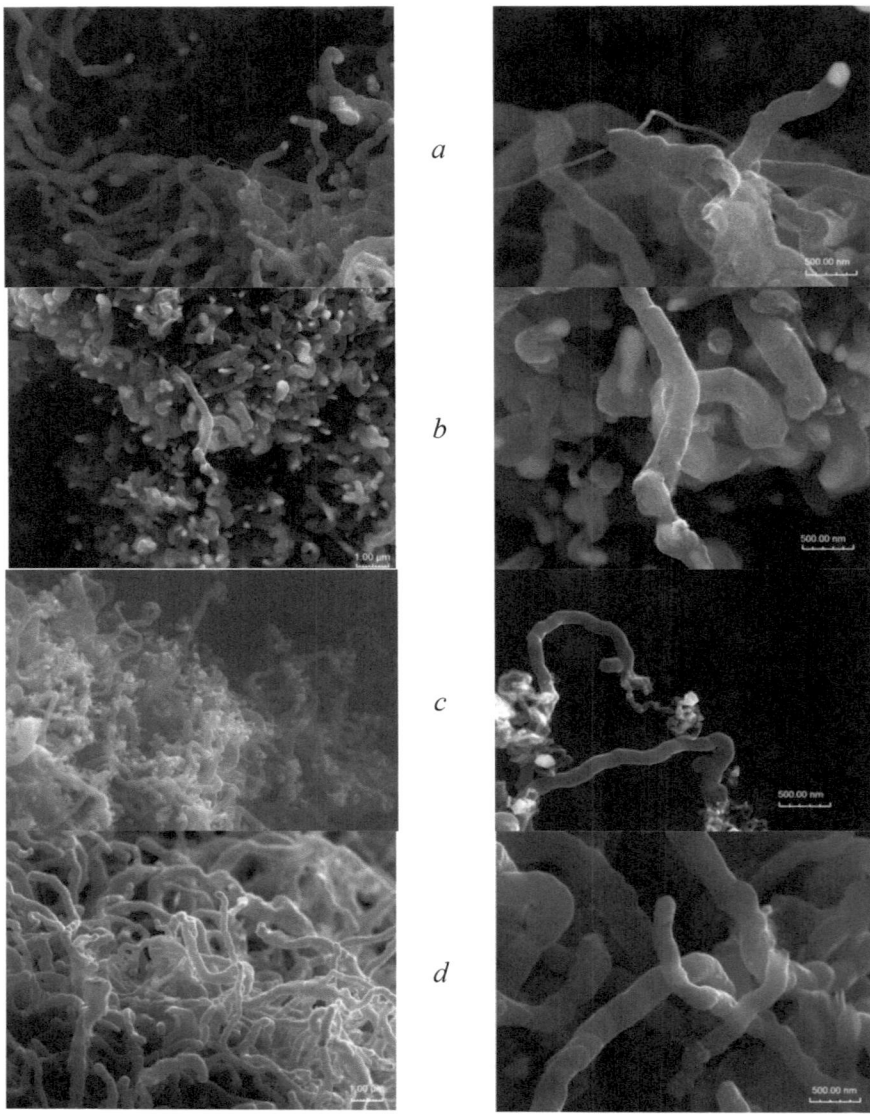

Fig. 1.38 Morphology of carbon nanomaterials synthesized on steel plates depending on their preliminary processing: **a**, **d**—two-step oxidation; **b**, **c**—one-step oxidation

valence tangential carbon–carbon vibrations and characterizes the degree of order of the carbon phase. The ratio of these two modes allows determining the number of defects in the nanotubes. In the case of a not well-ordered graphite, Raman spectral lines are observed in the ranges of 1585–1570 and 1350–1300 cm^{-1} [101].

Raman spectra in Fig. 1.39 are practically identical. The D and G characteristic modes at 1300 and 1590 cm^{-1} are observed; they can correspond to single-walled nanotubes. The D mode has a higher intensity compared to other samples, which may indicate slight bending of the nanotubes and higher density of defects. The ratios of the areas under the G and D bands were calculated (Table 1.4). These ratios are quantitative characteristics of the density of defects in single-walled nanotubes. The least defective is the sample (a) (Table 1.4).

Summary

CNTs were synthesized from methane air conversion products by chemical deposition from gas in a flow reactor. It is shown that the synthesis of CNTs at 923 K increases the mobility of catalyst particles, which leads to their agglomeration, and

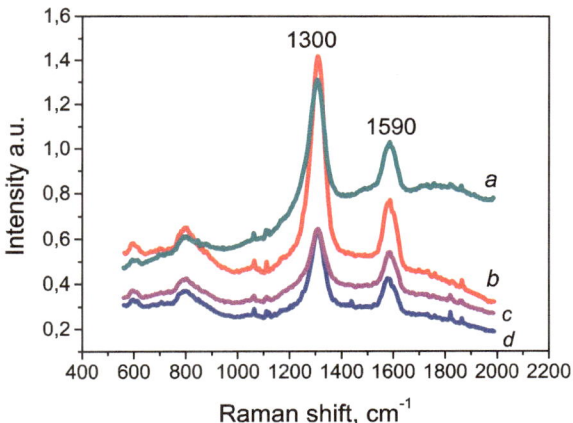

Fig. 1.39 Raman spectra of carbon nanomaterials

Table 1.4 Ratio of areas under G and D bands in the deposited carbon materials

Sample according to (Fig. 1.39)	$X = \frac{I_D}{I_G}$, arb. units	Area ratio, arb. unit	
a	1.906	$S(D)$	238.56
		$S(G)$	125.14
b	1.938	$S(D)$	119.16
		$S(G)$	61.5
c	1.957	$S(D)$	89.34
		$S(G)$	45.65
d	2.647	$S(D)$	210.41
		$S(G)$	79.49

so affects the growth of CNTs. The basis of the technology for the synthesis of carbon materials is developed, which allows obtaining CNTs with different morphologies and thicknesses. The structure of the CNTs and its dependence on the conditions of synthesis were studied using scanning electron microscopy.

Raman spectra of the carbon nanotubes contain two characteristic peaks at 1300 and 1590 cm^{-1}, which correspond to the D and G modes, respectively. The ratios of the areas under the G and D bands were calculated, which indicated a significant number of defects in the carbon nanotubes.

References

1. Mischenko SV, Tkachev AG (2008) Carbon nanomaterials. Fabrication, properties and applications. Mechanical Engineering, Moscow, 320 p
2. Rostovtsev ST (1956) Theory of metallurgical processes. Metallurgizdat, Moscow, 515 p [in Russian]
3. Gomzikov AI (2005) Structural and texture formation in electrotechnical anisotropic steel made using the process of nitriding. Ph.D., UGTN, Yekaterinburg, 19 p
4. Bondarenko BI, Shapovalov VA, Garmash NI (2003) Teorija i tehnologija beskoksovoj metallurgii. Nauk. Dumka, Kiev, 536 p
5. Rojter VA (1968) Kataliticheskie svojstva veshhestv. Spravochnik. Naukova Dumka, Kiev, 1464 p
6. Rostovcev ST (1956) Teorija metallurgicheskih processov. Metallurgizdat, Moscow, 515 p
7. Esin OA, Gel'd PV (1962) Fizicheskaja himija pirometallurgicheskih processov. Metallurgizdat, Sverdlovsk, 703 p
8. Kolesnik NF, Kudievskij SS, Kirichenko AG, Priluckij OV (2006) Termokataliticheskij raspad monooksida ugleroda. Izdatel'stvo Zaporozhskoj gosudarstvennoj inzhenernoj akademii, Zaporozh'e, 336 p
9. Chesnokov VV, Bujanov RL (2000) Obrazovanie uglerodnyh nitej pri katalitricheskom razlozhenii uglevodorodov na metallah podgruppy zheleza I ih splavah. Usp himii 69(7):675–692
10. Pod red. Gnesina GG, Skorohoda VV (2008) Neorganicheskoe materialovedenie. Osnovy nauki o materialah. T.1. Naukova Dumka, Kiev, 1152 p
11. Derjagin BV, Fedoseev DV (1977) Rost almaza i grafita iz gazovoj fazy. Nauka, Moscow, 116 p
12. Messerle VE, Ustimenko AB (2013) Plazmennaja pererabotka uglej. Teploenergetika 12:23–28
13. Bondarenko B, Sviatenko O, Kotov V, Khovavko A, Filonenko D, Nebesniy A, Vishnevsky A (2013) Research of the technology of carbon nanotubes production in gas mixtures contained CO. Int J Energy Clean Environ 14(2–3):177–182
14. Khovavko A, Sviatenko A, Kotov V, Bondarenko B, Nebesniy A, Filonenko D (2013) Technology of carbon nanotubes in gas mixtures containing carbon monoxide. Phys Status Solidi 10(7–8):1180–1182
15. Uvarova IV (1990) Fenomenologicheskie aspekty dispergirovanija produktov pritopohimicheskih reakcijah vosstanovlenija metallov iz oksidov (obzor). Poroshkovaja metall (2):59–65
16. Livshic BG (1971) Metallografija. Metallurgija, Izd. 2-e, ispr. i dop, Moscow, 405 p
17. Guljaev AP (1978) Metallovedenie. Metallurgija, Moscow, 647 p
18. Pod red. Kana R (1968) Fizicheskoe metallovedenie. Vyp. 2. Fazovye prevrashhenija. Metallografija. Per. s angl. Mir, Moscow, 490 p
19. Bokshtejn BS, Bokshtejn SZ, Zhuhovickij AA (1974) Termodinamika i kinetka diffuzii v tverdyhtelah. Metallurgija, Moscow, 280 p

20. Dzhejkok M, Parfit Dzh (1984) Himija poverhnostej razdela faz. Mir, Moscow, 269 p
21. Korotich V (1978) Osnovy teorii i tehnologii podgotovki syr'ja k domennoj plavke. Metallurgija, Moscow, 206 p
22. Harris P (2003) Uglerodnye nanotruby i rodstvennye struktury. Novyematerialy XXI veka. Tehnosfera, Moscow, 336 p
23. Pikunov MV, Desipri AI (1980) Metallovedenie. Metallurgija, Moscow, 256 p
24. Kurdjumov GV, Utevskij LM, Jentin RI (1977) Prevrashhenie v zheleze i stali. Nauka, Moscow, 238 p
25. Krishtal MA (1972) Mehanizm diffuzii v zheleznyh splavah. Metallurgija, Moscow, p 399
26. Vladimirov LP (1970) Termodinamicheskie raschety ravnovesija metallurgicheskih reakcij. Metallurgija, Moscow, p 528
27. Kotov VG, Svyatenko AM, Khovavko AI, Nebesnyi AA, Filonenko DS (2014) Eksperimental'nye issledovanija processa sazheobrazovanija pri vysokoj koncentracii vodoroda v gaze, soderzhashhem monooksid ugleroda. Energotehnol resursosberezhenie 2:33–38
28. Dubrov NF, Lapkin NI (1963) Elektrotehnicheskie stali. Metallurgija, Moscow, 384 p
29. Vashhenko AI, Zen'kovskij AG (1972) Okislenie i obezuglerozhivanie stali. Metallurgija, Moscow, 336 p
30. Bogdandi L, Jengel' GJ (1971) Vosstanovlenie zheleznyh rud. Metallurgija, Moscow, 520 p
31. Jestrin BM (1973) Proizvodstvo i primenenie kontroliruemyh atmosfer. Metallurgija, Moscow, 392 p
32. Naumov VA, Pavlova TN (1972) Issledovanie kinetiki reakcii disproporcionirovanija okisi ugleroda na zheleznom katalizatore. ZhFH 6:1480–1483
33. Ivensen VA (1985) Fenomenologija spekanija i nekotorye voprosy teorii. Metallurgija, Moscow, 246 p
34. Tkachev A, Mishhenko S, Negrov V (2007) Industrial production of carbon nanostructured material "Taunit". Nanoindustrija (2):24–26 (Rus.)
35. Nebesniy AA, Kotov VG, Svyatenko MA, Filonenko DS, Khovavko AI, Bondarenko BI (2015) Carbon. Nanomaterial formation at the processing of fresh-reduced iron by products of natural gas conversion. Energy Technol Recour Sav (5–6):34–42 (Rus.)
36. Shumilina ZF, Jagodkin VI, Shpoljanskij MA, Aleksandrov SV, Fedikjun JT, Dronova NN (1975) An experimental study of steam reforming of natural gas under pressure up to 100 atm. In: Catalytic conversion of hydrocarbons, issue 2. Naukova Dumka, Kiev, pp 9–12 (Rus.)
37. Bondarenko BI, Bezuglij VK (2002) Potentials of components of physico-chemical systems. Academiperiodica, Kyiv, 125 p (Ukr.)
38. Shmykov AA, Malyshev BV (1953) Controlled atmospheres during steel thermal treatment. MAShGIZ, Moscow, 372 p (Rus.)
39. Kasatkin AG (1973) Basic processes and apparatuses of chemical technology. Chemistry, Moscow, 750 p (Rus.)
40. Temkin MI, Shub FS, Homchenko AA, Apel'baum LO (1977) Kinetics of methane conversion on a nickel catalyst. In: Scientific bases of catalytic conversion of hydrocarbons. Naukova Dumka, Kiev, pp 3–27 (Rus.)
41. Sharifov AP, Zhidkov BA (1979) Investigation of the kinetics of methane conversion reaction with water vapor. In: Catalytic conversion of hydrocarbons, issue 4. Naukova Dumka, Kiev, pp 65–69 (Rus.)
42. Pomerancev VM, Anufrieva TA (1980) Investigation of methane conversion process on nickel catalyst. In: Catalytic conversion of hydrocarbons, issue 5. Naukova Dumka, Kiev, pp 14–17 (Rus.)
43. Kopytov VF (1955) Steel heating in the furnaces. Metallurgizdat, Moscow, 264 p (Rus.)
44. Kotov VG, Sviatenko AM, Filonenko DS, Nebesniy AA (2012) The prediction of reducing gas specific consumption for iron direct production in shaft furnaces. Energy Technol Recour Sav (6):39–43 (Rus.)
45. Jestrin BM (1963) Production and application of controlled atmospheres. Metallurgiya, Moscow, 343 p (Rus.)

46. Atroshhenko VI, Loboiko AJ, Jurchenko AP, Zviagincev GL (1977) Study of kinetics of methane and carbon monoxide conversion of under pressure. In: Scientific bases of catalytic conversion of hydrocarbons. Naukova Dumka, Kiev, pp 51–62 (Rus.)
47. Kotov VG, Sviatenko AM, Khovavko AI, Nebesniy AA, Filonenko DS (2014) Thermodynamics of carbon-black formation process at high hydrogen concentration in gas which contains carbon monoxide. Energy Technol Recour Sav (2):33–38 (Rus.)
48. Kotov VG (1980) Transition period of sintering process of sinter charge with recirculation of exhaust gases. Izv. vuzov Chernajametallurgija (10):25–29
49. Under the red of Leibush AG (1971) Production of technological gas for the synthesis of ammonia and methanol from hydrocarbon gases. Khimiya, Moscow, 286 p (Rus.)
50. Bondarenko BI, Sviatenko AM, Savenko LV (1969) Carburizing of iron-ore materials at their reduction in fluidized bed by converted gas. In: The use of natural gas in the industry. Naukova Dumka, Kiev, pp 4–8
51. Bondarenko B, Khovavko O, Bogomolov V, Kozhan O, Nazarenko V, Simeyko K (2012) Microspark plasma capsulation of quartz sand by continuous deposition of pyrolitic carbon nanolayers. In: Theses and reports 4th international conference "NANOCON 2012", Brno, Czech Republic, EU, p 1
52. Bondarenko B, Bogomolov V, Kozhan A, Khovavko A, Nazarenko V, Simeyko K (2013) Development of technological foundations for pure silicon production by carbothermic reduction. Int J Energy Clean Environ 14(2–3):183–189
53. Strativnov EV (2015) Design of modern reactors for synthesis of thermally expanded graphite. Nanoscale Res Lett 10:245
54. Strativnov EV (2017) Unit with fluidized bed for gas-vapor activation of different carbonaceous materials for various purposes: design, computation, implementation. Nanoscale Res Lett 12:122
55. Vavrysh AS, Nebesniy AA, Bondarenko BI, Khovavko AI, Snigur AV, Sviatenko OM (2013) Production of carbon nanomaterials for hydrogen storage. In: Theses and reports 5th international conference "NANOCON 2013", Brno, Czech Republic, EU
56. Danafar F, Fakhru'lRazi A, Salleh MAM, Biak DRA (2009) Fluidized bed catalytic chemical vapor deposition synthesis of carbon nanotubes. A review. Chem Eng J 155:37–48
57. Pylypenko PS (1969) Kinetics of carbonization of iron, cobalt and nickel by methane. In: The use of natural gas in the industry. Naukova Dumka, Kiev, pp 212–218
58. Medvedeva LI, Evstafiev VC (1990) Catalytic carbonization of sponge iron with natural gas. In: All-Union scientific and technical conference "Problems of the theory and technology of iron ore preparation for the blast furnace process and non-coking metallurgy". Dnipropetrovsk, p 15
59. Roiter VA (1968) Catalytic properties of substances. Directory. Naukova Dumka, Kiev, 1462 p
60. Bondarenko BI, Bezugly VK (2002) Potentials of components of physical and technical systems. Akademperiodika, Kiev, 125 p
61. Silva RCF, Ardisson JD, Cotta AAC, Araujo MH, de Carvalho Teixeira AP (2020) Use of iron mining tailings from dams for carbon nanotubes synthesis in fluidized bed for 17α-ethinylestradiol removal. Environ Pollut 260(114099):1–9. https://doi.org/10.1016/j.envpol. 2020.114099
62. Almkhelfe H, Carpena-Nunez J, Back TC, Amama PB (2016) Gaseous product mixture from Fischer–Tropsch synthesis as an efficient carbon feedstock for low temperature CVD growth of carbon nanotube carpets. Nanoscale 8(27):13476–13487. https://doi.org/10.1039/C6NR03 679A
63. Pierson HO (1999) Handbook of chemical vapor deposition: principles, technology and applications. William Andrew Publishing, Elsevier, Norwich
64. Shah KA, Tali BA (2016) Synthesis of carbon nanotubes by catalytic chemical vapour deposition: a review on carbon sources, catalysts and substrates. Mater Sci Semicond Process 41:67–82. https://doi.org/10.1016/j.mssp.2015.08.013

65. Jagadeesan AK, Thangavelu K, Dhananjeyan V (2020) Carbon nanotubes: synthesis, properties and applications. In: Pham P (ed) 21st century surface science—a handbook. IntechOpen, London, pp 1–21. https://doi.org/10.5772/intechopen.92995

66. Huang ZP, Wang DZ, Wen JG, Sennett M, Gibson H, Ren ZF (2002) Effect of nickel, iron and cobalt on growth of aligned carbon nanotubes. Appl Phys A 74(3):387–391. https://doi.org/10.1007/s003390101186

67. Haris PJF (2009) Carbon nanotubes science: synthesis properties and applications. Cambridge University Press, Cambridge

68. Khovavko O, Nebesnyi A, Filonenko D, Barabash M, Leonov D, Svyatenko O (2022) Peculiarities of the formation of carbon nanotubes from the products of the air conversion of methane by the method of chemical deposition. Nanosyst Nanomater Nanotechnol 20(3):715–724. https://doi.org/10.15407/nnn.20.03.715

69. Min YK, Eom T, Kim H, Kang D, Lee S-E (2023) Independent heating performances in the sub-zero environment of MWCNT/PDMS composite with low electron-tunneling energy. Polymers 15(5):1171. https://doi.org/10.3390/polym5051171

70. Rozhin P, Kralj S, Soula B, Marchesan S, Flahaut E (2023) Hydrogels from a self-assembling tripeptide and carbon nanotubes (CNTs): comparison between single-walled and double-walled CNTs. Nanomaterials 13(5):847. https://doi.org/10.3390/nano13050847

71. Kamedulski P, Zielinski W, Nowak P, Lukaszewicz JP, Ilnicka A (2020) 3D hierarchical porous hybrid nanostructure of carbon nanotubes and N-doped activated carbon. Sci Rep 10(1):1–11. https://doi.org/10.1038/s41598-020-75831-x

72. Kamedulski P, Lukaszewicz JP, Witczak L, Szroeder P, Ziolkowski P (2021) The importance of structural factors for the electrochemical performance of graphene/carbon nanotube/melamine powders towards the catalytic activity of oxygen reduction reaction. Materials 14(9):2448. https://doi.org/10.3390/ma14092448

73. Khovavko A, Sviatenko A, Kotov V, Bondarenko B, Nebesniy A, Filonenko D (2013) Technology of carbon nanotubes production in gas mixtures containing carbon monoxide. Phys Status Solidi C 10(7–8):1180–1182. https://doi.org/10.1002/pssc.201200971

74. Nebesnyi A, Kotov V, Sviatenko A, Filonenko D, Khovavko A, Bondarenko B (2017) Carbon nanomaterial formation on fresh-reduced iron by converted natural gas. Nanoscale Res Lett 12(107):1–7. https://doi.org/10.1186/s11671-017-1882-6

75. Cassell AM, Raymakers JA, Kong J, Dai H (1999) Large scale CVD synthesis of single-walled carbon nanotubes. J Phys Chem B 103:6484–6492

76. Cheng HM, Su F, Li G, Pan HY, He LL, Sun X, Dresselhaus MS (1998) Large-scale and low-cost synthesis of single-walled carbon nanotubes by the catalytic pyrolysis of hydrocarbons. Appl Phys Lett 72:3282–3284

77. Li Y, Zhang X, Shen L, Luo J, Tao X, Liu F, Xu G, Wang Y, Geise HJ, Van Tendeloo G (2004) Controlling the diameters in large scale synthesis of single-walled carbon nanotubes by catalytic decomposition of CH_4. Chem Phys Lett 398:276–282

78. Nebesnyi A, Kotov V, Sviatenko A et al (2017) Carbon nanomaterial formation on fresh-reduced iron by converted natural gas. Nanoscale Res Lett 12:107

79. Keller TM, Laskoski M, Qadri SB (2007) Solid-phase synthesis of multi-walled carbon nanotubes from butadiynyl-ferrocene-containing compounds. Carbon 45(2):443–448. https://doi.org/10.1016/j.carbon.2006.08.014

80. Zambri MSM, Mohamed NM, Kait CF (2011) Preparation of electrochromic material using carbon nanotubes (CNTs). J Appl Sci 11(7):1321–1325

81. Moisalaa A, Nasibulina AG, Browna DP, Jiangb H, Khriachtchevc L, Kauppinena EI (2006) Single-walled carbon nanotube synthesis using ferrocene and iron pentacarbonyl in a laminar flow reactor. Chem Eng Sci 61(13):4393–4402

82. Wulan PPDK, Setiawati NS (2018) The effect of mass ratio of ferrocene to camphor as carbon source and reaction time on the growth of carbon nanotubes. E3S Web Conf 67:03037

83. Kumar A, Sharma K, Dixit AR (2020) Carbon nanotube- and graphene-reinforced multiphase polymeric composites: review on their properties and applications. J Mater Sci 55(3):2682–2724

84. Herrera JE, Balzano L, Borgna A, Alvarez WE, Resasco DE (2001) Relationship between the structure/composition of Co-Mo catalysts and their ability to produce single-walled carbon nanotubes by CO disproportionation. J Catal 204(1):129–145
85. Rakov EG (2007) Preparation of thin carbon nanotubes by catalytic pyrolysis on a support. Russ Chem Rev 76(1):3–26. https://doi.org/10.1070/RC2007v076n01ABEH003641(Rus.)
86. Shanov V, Yun Y-H, Schulz MJ (2006) Synthesis and characterization of carbon nanotube materials (review). J Univ Chem Technol Metall 41(4):377–390
87. Teo KBK (2003) Catalytic synthesis of carbon nanotubes and nanofibers. In: Encyclopedia of nanoscience and nanotechnology, vol 10, pp 1–22
88. Maskaeva LN, Fedorova EA, Markova VF (2019) Thin films and coatings technology. Study guide. Ministry of Science and Higher Education R.F., Ural Federal University. Publishing House of the Ural University, Yekaterinburg, 236 p. ISBN 978-5-7996-2560-3 (Rus.)
89. Luchkin AG, Luchkin GS (2012) Surface cleaning of substrates for coating by vacuum-plasma methods. Bulletin of Kazan Technological University, Kazan, pp 208–210 (Rus.)
90. Barabash MY, Grinko DO, Sperkach SO (2015) Formation of nanostructure on templates by radiation from a visible shade. IMFNASU, Kyiv, 202 p (Ukr.)
91. Bondarenko BI, Sviatenko OM, Khovavko AI, Kotov VG, Nebesnyi AA, Filonenko DS (2018) Big multi-walled carbon nanotubes synthesis using a reduced iron as a catalyst. In: Proceedings of the 2018 IEEE 8th international conference on nanomaterials: applications & properties (NAP-2018), Zatoka, Odessa Region, Ukraine, 9–14 Sept 2018, pp 50–53. https://doi.org/10.1109/NAP.2018.8914865
92. Słoczyński J, Janas J, Machej T, Rynkowski J, Stoch J (2000) Catalytic activity of chromium spinels in SCR of NO with NH$_3$. Appl Catal B Environ 24:45–60
93. Chesnokov VV, Buyanov RA (2000) The formation of carbon filaments upon decomposition of hydrocarbons catalysed by iron subgroup metals and their alloys. Russ Chem Rev 69(7):623–638. https://doi.org/10.1070/RC2000v069n07ABEH000540
94. Mishchenko SV, Tkachev AG (2008) Carbon nanomaterials. Production, properties and applications. Mashinostroenie, Moscow, 320 p (Rus.)
95. Harris PJF (2007) Solid state growth mechanisms for carbon nanotubes. Carbon 45:229–239. https://doi.org/10.1016/j.carbon.2006.09.023
96. Yanchuk IB, Koval's'ka EO, Brichka AV, Brichka SY (2009) Raman scattering studies of the influence of thermal treatment of multi-walled carbon nanotubes on their structural characteristics. Ukr J Phys 54(4):407
97. Kumar A, Sharma K, Dixit AR (2020) Carbon nanotube- and graphene-reinforced multiphase polymeric composites: review on their properties and applications. J Mater Sci 55:2682. https://doi.org/10.1007/s10853-019-04196-y
98. Bondarenko BI, Sviatenko OM, Khovavko AI, Kotov VG, Nebesnyi AA, Filonenko DS (2018) Big multi-walled carbon nanotubes synthesis using a reduced iron as a catalyst. In: Proceedings of the 2018 IEEE 8th international conference on nanomaterials: applications & properties (NAP-2018), Zatoka, Odesa Region, Ukraine, 9–14 Sept 2018, pp 50–53. https://doi.org/10.1109/NAP.2018.8914865
99. Lupis K (1989) Chemical thermodynamics of materials. Metallurgiya, Moscow (Russian translation)
100. Mishchenko SV, Tkachev AG (2008) Carbon nanomaterials: production, properties, applications. Mashinostroenie, Moscow (in Russian)
101. Yanchuk IB, Kovalska EO, Brichka AV, Brichka SY (2009) Raman scattering studies of the influence of thermal treatment of multi-walled carbon nanotubes on the their structural characteristics. Ukr J Phys 54(4):407. http://archive.ujp.bitp.kiev.ua/files/journals/54/4/540412p.pdf

Chapter 2
Design of Modern Equipment
for Synthesis of Carbon Nanomaterials

Abstract The results of experimental studies aimed to multilayer graphene obtaining. Thermoexpanded graphite was used as a raw material. The technological regulations consisted of two main stages: obtaining a water-graphite suspension by cavitation method and grinding graphite globules on graphene plates using ultrasound or electrohydraulic impact according to Yutkin's method. As a result of the experiment, a few layers of graphene were obtained, which is a pack of 10–20 graphene plates in each. CFD modeling of the cavitation process was performed. The analysis showed that the chosen geometry of the impeller blades, its speed, and angle of attack provides the necessary effect—cavitation and the necessary mixing throughout the volume. Graphite-graphene composites (GGC) have been obtained as a result of mechanical treatment of thermoexpanded graphite (TEG).

2.1 Design of Modern Reactors for Synthesis of Thermally Expanded Graphite (TEG)

One of the most progressive trends in modern science and technology is the development of energy-efficient technologies for synthesis of nanomaterials. Nanolayered graphite (thermally expanded graphite) is one of the key nanomaterials based on carbon. Due to its unique properties (chemical and thermal stability, ability to form without a binder, elastic, etc.), it can be used as an effective absorber of organic substances, the material for seals manufacturing for such important industries as gas transportation, automobile, etc. Thermally expanded graphite is a promising material for hydrogen and nuclear energy industry. The development of thermally expanded graphite production is resisted by high specific energy consumption at its manufacturing and by some technological difficulties. Therefore, the creating of energy-efficient technology for its production is a very actual point.

The Stages of TEG Obtaining. The raw material for TEG is natural crystalline flake graphite (Fig. 2.1). Deposits of flake graphite are situated in Ukraine (Zavalie, Kirovograd region), China, Russia, and the USA. The great manufacturers of graphite are China, India, Korea, and Brazil. At the first stage, the natural crystalline graphite is

© The Author(s), under exclusive license to Springer Nature Switzerland AG 2024   69
A. Khovavko et al., *Carbon Nanostructured Materials*,
SpringerBriefs in Materials, https://doi.org/10.1007/978-3-031-64121-3_2

crushed into flakes and oxidized [1, 2]. Oxidation is an implementation of molecules and ions of sulfuric or nitric acid between the layers of the graphite's crystal lattice in the presence of an oxidant (hydrogen peroxide, potassium permanganate, and others). This process is called intercalation. This material is called oxidized graphite (OG).

Then OG is washed by water to remove excess of intercalant and dried to give it the technological properties (Fig. 2.2).

The next step is OG heating in the high-temperature reactor to the temperature of OG expansion. Due to the extremely high heating rate (600 °C/s), there is a sharp expansion of gaseous products of decomposition of introduced sulfuric acid from the graphite's crystal lattice. Thermally expanded graphite is obtained by increasing the interlayer distance in the scaly graphite (about 300 times). A view of TEG at the micro and macrolevels is shown in Figs. 2.3 and 2.4.

A separate part of TEG is a piece of worm-like shape with a diameter of 0.1–0.5 mm and a length of 6–10 mm. This particle consists of a set of interconnected individual layers of graphite (graphene) and packs (10–50 PCs) of such layers.

Fig. 2.1 View of natural crystalline graphite

Fig. 2.2 View of intercalated flake graphite

Fig. 2.3 View of the TEG on the macrolevel

Fig. 2.4 View of the TEG on the microlevel

General Use of TEG

Gaskets and seals can be manufactured from TEG (Fig. 2.5) with unique properties by pressing and rolling without any binders and reinforcement [3, 4]. Products can be used in a wide temperature range and under the influence of aggressive environments. Thus, sealing goods found their application in various engineering industries—from cryogenic technologies to nuclear. TEG is used as a component in the batteries of new generation [5]. In the case of TEG addition to the cathode active mass of alkaline

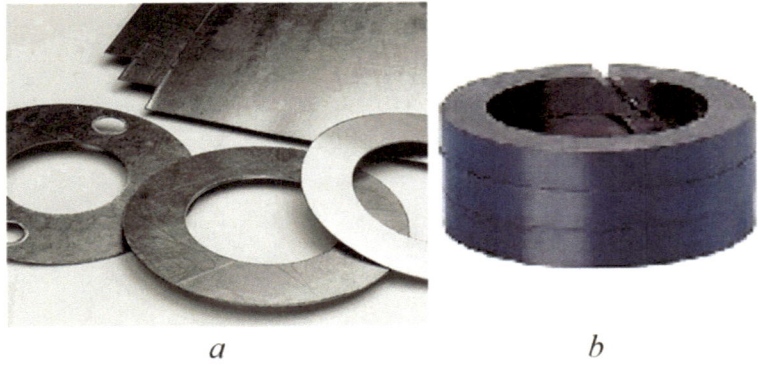

a   *b*

Fig. 2.5 Products from TEG by rolling (**a**) and pressing (**b**)

Zinc-Manganese Dioxide batteries, the coefficient of utilization of active material reaches up to 40% versus 29% in the case of application of natural graphite.

TEG is also a unique absorbent of such substances as helium, argon, nitrogen, krypton, hydrogen, xenon, isooctane, benzene, and cyclohexane. Basically, TEG is used as the absorber [6] for organic substances (Fig. 2.6). Due to its nature, TEG absorbs liquid organics not only by individual particles, but by clusters of its particles.

Heating Methods of Oxidized Graphite (OG) Particles. OG heating by heat from combustion of liquid or gaseous fuel currently is the most common way. It is realized in the furnace with a fluidized bed of inert heat carrier and in furnaces with a cocurrent flow of flue gases with the target product. This technology was taken as the basis for a comprehensive study and analysis. The most important factor which affects to TEG quality and energy consumption is the heating rate of OG. For intensification of heat exchange between OG particles and the flow of fuel combustion products, it increased the turbulence in the reaction zone. Also OG feeding was accomplished directly into the core of the flame [7]. Specially developed devices have complex geometry of

Fig. 2.6 Use of TEG for water purification in wastewater treatment plants

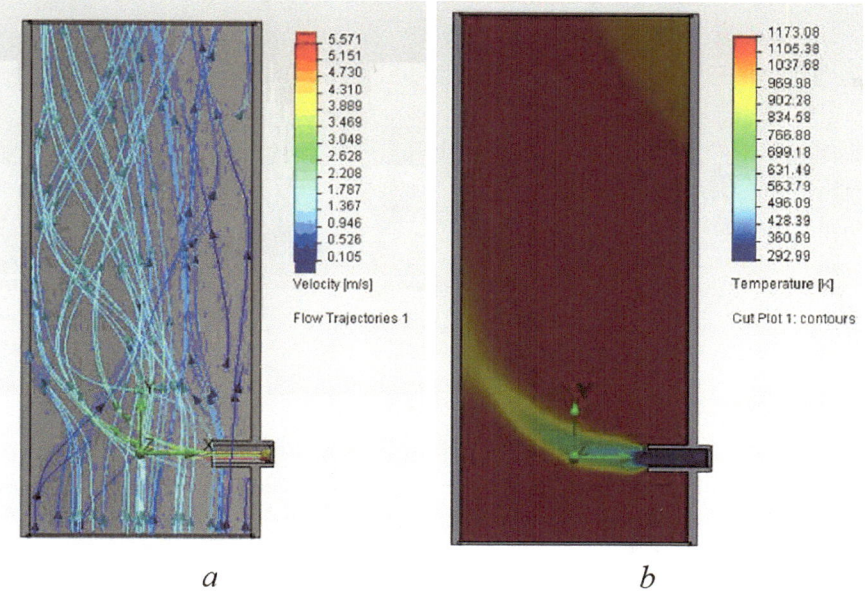

Fig. 2.7 Gas flows (**a**) and temperature in the reactor (**b**) for TEG synthesis by "the classical scheme"

the internal space (and not symmetric). Therefore, modeling of physical processes occurring inside of such devices it is proposed to carry out by 3D with the help of modern software systems (for example, ANSYS and SolidWorks) [8, 9]. In this manuscript, we considered two new type of the reactors: reactor with two opposite burners (one of which—"burner-feeder") and "vortex-type" apparatus. Feeding of combustion products and OG into the reaction zone is carried out tangentially.

The significant deficiency of the "classic" technological scheme was identified while analyzing of the hydrodynamics of the flow and temperature fields distribution inside the reactor. This disadvantage consists at the following. Cold air is fed with raw materials into the reactor and dilutes by itself smoke gases and thereby lowers the temperature in the reaction zone (gas flows and their temperature are shown in Fig. 2.7). Flue gases with a higher temperature are not used it is not rational.

2.2 Choosing the Type of Apparatus and Calculation of Material Flows

It is shown that to improve TEG quality—it is necessary to increase OG heating rate. As a result, we created a new technology for TEG production. The main idea of proposed method is that the raw materials are fed into the reactor together with air,

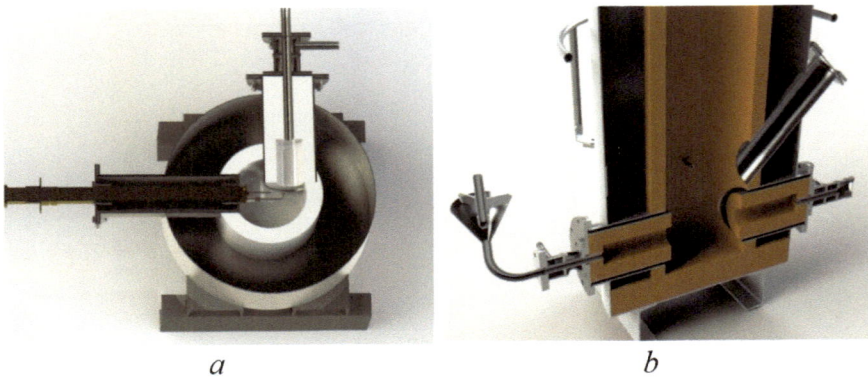

a b

Fig. 2.8 Reactors for the production of TEG: **a** "cyclone type", **b** "opposite type"

used for fuel combustion. Moreover, the gas-dynamic characteristics of two-phase flow are that allow OG particles to deposit directly into the core of the flame. OG particles are actively heated in the core of the torch, effectively using of convective and radiant component of heat transfer. High rate of convection component is implemented due to the reactor's zone of turbulence and by the dynamic pulsations that occur while fuel combustion. And the level of radiant component occurs due to the direct contact of "glowing" flame and OG particles.

New technological scheme is implemented in reactors: the reactor of "opposite type" and the reactor of "cyclone type" (Fig. 2.8).

In "cyclone-type" reactor, the flow of combustion products is fed tangentially. Thanks to such feeding scheme the product's residence time inside the reactor increases, while the dimensions of the device are reduced. The expanding occurs in the burner-feeder; at the same time, the vertical section of the reactor is used for the final pre-expansion and additional annealing of TEG in order to remove residual compounds of intercalant. Increasing the heating rate and the total reaction temperature allowed getting TEG of higher quality with a density of 3 g/l and higher purity of the product. These can extend the scope of application of this material up to its application in the nuclear industry.

The design of the apparatus involves the calculation of its geometrical parameters. This calculation means the determination of its material flows. The calculations are carried out basing on the hover speed of OG particles (for OG pneumatic transportation) and TEG (for TEG removal from the working area of the reactor). Calculation of material flows is connected with the volume of flue gases proceed from fuel combustion and the desired performance of the aggregate. According to these calculations, the source material (OG) must be guaranteed to be delivered in a high-temperature zone of the reactor and the finished product (TEG) has to leave it due to the speeds of the appropriate gaseous flows.

2.3 Study of Hydrodynamics in Reactors of New Type

New technology has allowed a much better use of the heat from fuel combustion, which had a positive impact on the values of specific energy consumption. In particular, for the apparatus of "opposite type" as it shown in Fig. 2.9 due to the active mixing of the gas streams at the bottom of the reactor, there was almost complete equalization of temperature throughout the volume in the vertical part of the apparatus. In this particular case, to illustrate the mixing process, cold air was fed.

This plot shows the data after the complete cessation of combustion in the burner-feeder. That is in the most unfavorable (emergency) time of apparatus operation. Operating mode (when both burners are working) knowingly provides the better gases mixing and temperature equalization in the reactor. The height and diameter of the reactor are designed to ensure the continuous (about 15 s) material staying in the working zone while the "working" regimes of apparatus. Longer stays of TEG in the high-temperature zone can additionally calcinate the final product. The size of the turbulence zone in the tunnel burner is designed for active mixing of hot gases and to form the cyclically flows (Fig. 2.10). This hydrodynamic regime provides for auto-ignition of feeding mixture and for combustion stabilization.

The eddy zone of the burner is identical to the eddy zone of the burner-feeder of the TEG generation apparatus, where direct contact of OG and hot gases is occurred.

Fig. 2.9 Mixing process of counter-flows in the reactor of opposite type

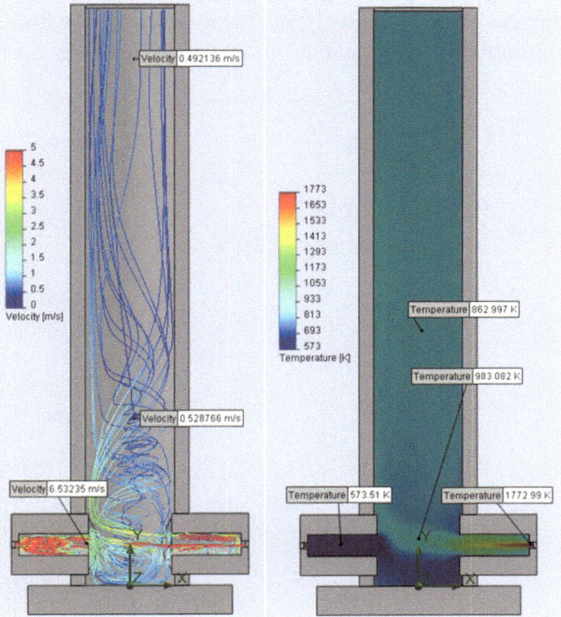

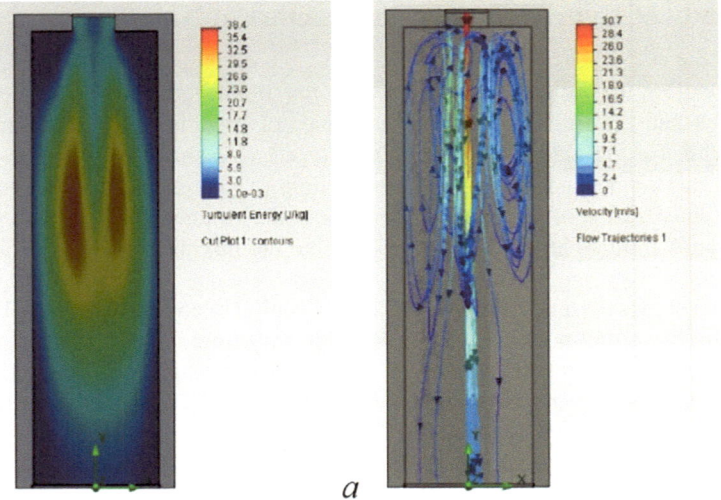

a b

Fig. 2.10 Trajectory, velocity of gas flows (**a**) and their corresponding distribution of turbulence energy (**b**) in the eddy zone of the burner

Turbulence energy in the eddy zone is 35–38 J/kg, which provides a significant (30-fold) increase of thermal conductivity of the flue gases (1.95 W/m K at 1000 °C) compared to their conductivity at the quiescent mode (0.0667 W/m K at 1000 °C). In the reactor of "cyclone type" due to swirling of a flow, TEG residence time increases at relatively small dimensions of the device (Fig. 2.11).

Fig. 2.11 Flows in the longitudinal (left) and transverse (right) cross sections of the cyclone-type aggregate

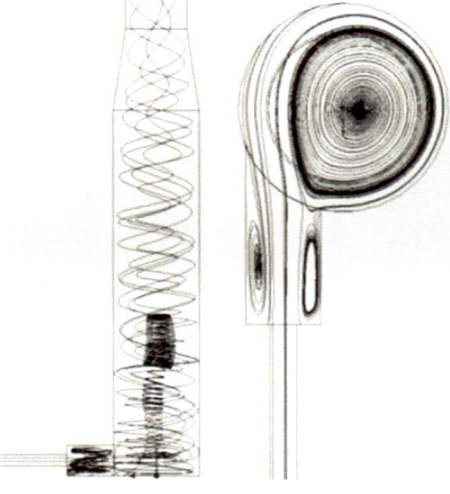

Simulation of Methane Combustion. The process of fuel combustion contributes additional ripples to flow hydrodynamics. In this regard, the heat exchange between OG particles and a flame has some nuances. Formal result of modeling of this process is the calculation of the final temperature and the heating rate of OG particles. However, while modeling process—big information massive and additional knowledge are appeared about behavior of the investigated material. So, for example, such parameters as: magnitude and the direction of particle trajectories, the Reynolds criterion for particles, time their staying in the reactor and other are need to be studied. All this information brings a researcher to the next step closer to the understanding of the processes occurring with the investigated material.

The simulation includes the several stages: construction of the geometry of the model, calculation grid construction, setting of source data, choosing of a solver, and actually solution of the task. To ensure the sufficient accuracy of the solution, the workspace in our case is divided into 1 million 28 thousand 519 cells of the grid (which corresponds to 1,473,273 nodes of a hexagonal grid).

Grid (Figs. 2.12 and 2.13) has a non-uniform break step to improve the accuracy of the calculation in the most "intense" sections of the space of the reaction zone. The grid is usually deliberately compacted in areas where there have been an increased in comparison with the rest space heat and mass transfer and also at the walls of the apparatus in order to account for near wall effects of hydrodynamic flow.

Fig. 2.12 Workspace of apparatus of a cyclone type

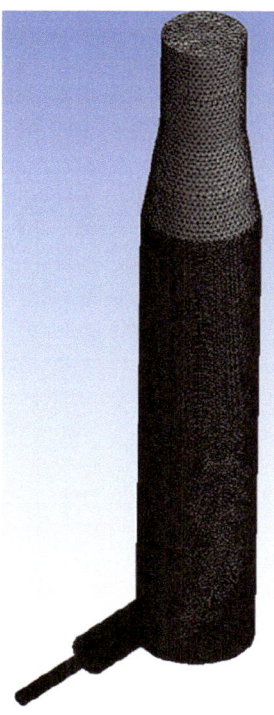

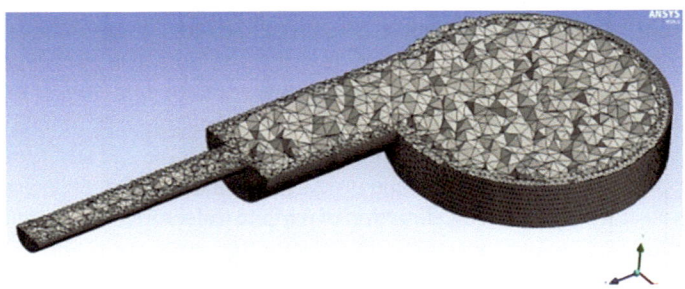

Fig. 2.13 Cross section of the active area of a cyclone-type apparatus

Modeling of the process of methane combustion is not the purpose of this work, but, because of its complexity, it is required a more detailed description, at least the original data and the applied models and methods. Author supposes that this information will be useful for professionals who engage in simulation by ANSYS. Table 2.1 presents the specific coefficients.

- The parameters of the solver which are used to simulate gas-thermal flow of methane combustion products;
- Turbulence model k-epsilon (2 eqn);
- Submodel—realizable;
- Option, considering the parietal effects—"near wall treatment" (standard wall function);
- Radiative transfer model—"P-1" model;
- Wavelength intervals: Band-0: 2.8–4.11; band-1: 4.5–8.76.

Table 2.1 Coefficients used for the design of process of burning of methane

No.	Reagents			Products			Arrhenius rate			Mixing rate	
		Stoich. coeff.	Rate exp		Stoich. coeff.	Rate exp	Pre-exp factor	Activation energy	Temperature exponent	A	B
1	CH_4	1	1.46	CO	1	0	1.66e15	1.7e8	0	4	0.5
	O_2	1.5	0.5217	H_2O	2	0					
2	CO	1	1.7	CO_2	1	0	7.9e14	9.6e7	0	4	0.5
	O_2	0.5	1.57								
3	CO_2	1	1	CO	1	0	2.2e14	5.2e8	0	4	0.5
				O_2	0.5	0					
4	N_2	1	0	NO	2	0	8.8e23	4.4e8	0	1e11	1e11
	O_2	1	4.01								
	CO	0	0.72	CO	0	0					
5	N_2	1	1	NO	2	0	9.27e14	5.7e8	− 0.5	1e11	1e11
	O_2	1	0.5								

Diagrams show the physical conditions of the flow of OG particles: the distribution of the temperature fields, the composition of combustion products, etc. Modern modeling tools and analysis (in this case ANSYS software) allow, also, operating with massive of OG particles of different size distribution, to determine their absolute and elative velocities, to determine the Reynolds criterion, etc. (Fig. 2.14).

Fig. 2.14 Plots of flow rates distribution, gas concentrations, levels of energy turbulence, and temperatures in the cross section of the cyclone-type reactor

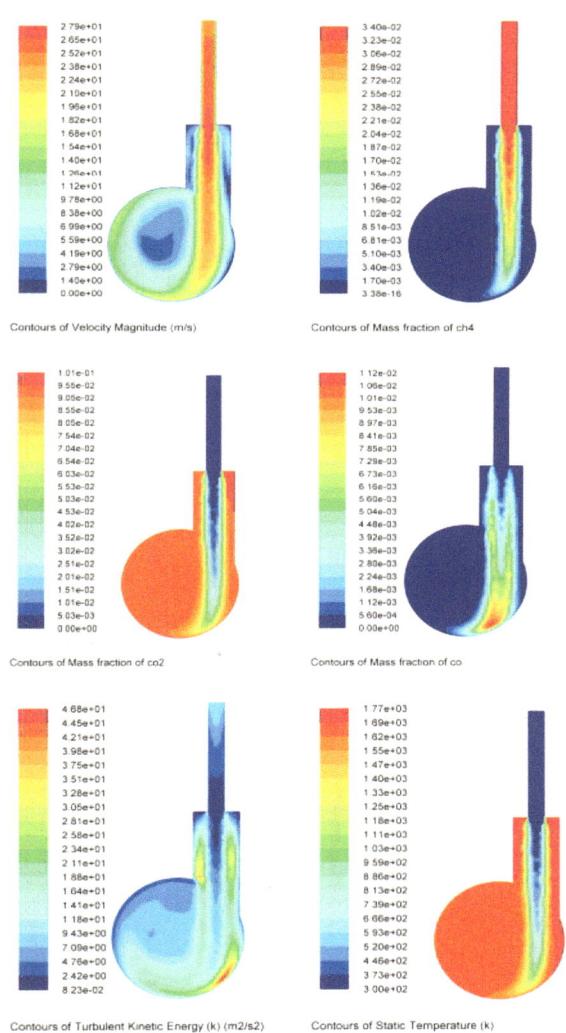

2.4 Investigation of the Heating Rate of Oxidized Graphite Particles in the New Type Reactor Considering the Hydrodynamic Pulsations that Occur Due to Methane Combustion

The oxidized graphite particles carried by a stream of hot products of combustion of methane was instantly heated up. To study the heating rate of the particles of OG in conditions as close as possible to the real, flow rates in the reactor are simulated (Fig. 2.15).

Solver's parameters used to simulate the behavior of OG particles in the flow of products of methane combustion:

- Volumetric
- Diffusion energy source
- Mixture material. Methane-air-2-step. Number of volumetric species $= 7$
- Properties of methane-air-2-step: Mixture species: CH_4, O_2, CO_2, CO, H_2O, NO, N_2
- Density—Incompressible ideal gas
- Heat capacity—Using the law of mixing environment
- Thermal conductivity—0.0241 W/m K
- Viscosity—1.72e−5 kg/ms

Fig. 2.15 Trajectory and temperature of OG particles in the "cyclone-type" reactor

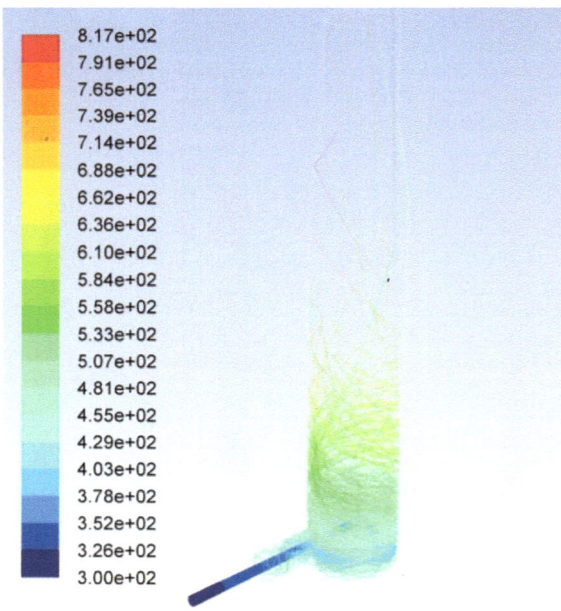

Particle Traces Colored by Particle Temperature (k)

- Mass diffusion—2.88e−5 m²/s
- Turbulence-chemistry interaction—Finite-rate/eddy dissipation
- The following forces, acting on OG particles in the gas flow, are considered at the simulation:
- Shape factor
- Particle radiation interaction
- Thermophoretic force
- Brownian motion
- Saffman lift force
- Virtual mass force
- Pressure gradient force
- The source data for OG particles:
- The equivalent diameter of the particles $D = 0.43$ mm
- Consumption of oxidized graphite $G = 30$ kg/h $= 0.0083$ kg/s
- The initial temperature of the particles $T = 300$ K
- The density $\rho = 2230$ kg/m³
- The heat capacity $Cp = 1680$ J/kg K
- The thermal conductivity of 0.33 W/m K.

2.5 Design of Modern Reactors for Thermoexpanded Graphite Synthesis

Modern design and construction of apparatus of a new type involve the several basic steps: the selection of a unit type, calculation of material flows, sizing of the reactor and of the diameters of all pipes, checking calculation of the rate of OG heating with the help of computer simulation, and finally—the designing by using of appropriate software. Connecting sizes of the reactor's details are chosen, if it is possible from standard range, to ensure seamless interfacing with adjacent equipment.

On the order of Argonne National Laboratory (USA), the highly efficient apparatus of the new type ("opposite type") for TEG synthesis was developed and manufactured at the Gas Institute of the National Academy of Sciences of Ukraine in 2013. In 2015, it is planned to introduce the next apparatus of the new type ("cyclone type") at one of the Ukraine's plant. Visualization of the new apparatus is presented in Fig. 2.16.

Summary

Modern requirements for energy efficiency and resource saving, as well as a decrease of weight of the industry aggregates, require more careful study of the processes occurring in the reactor and accurate calculations of the design and construction of units of a new type. For these purposes, it's appropriate to use the modern methods of numerical computer simulation at the field of computational fluid dynamics (CFD). Thus, energy efficiency during TEG generation and resource saving at equipment manufacturing for its production can be achieved.

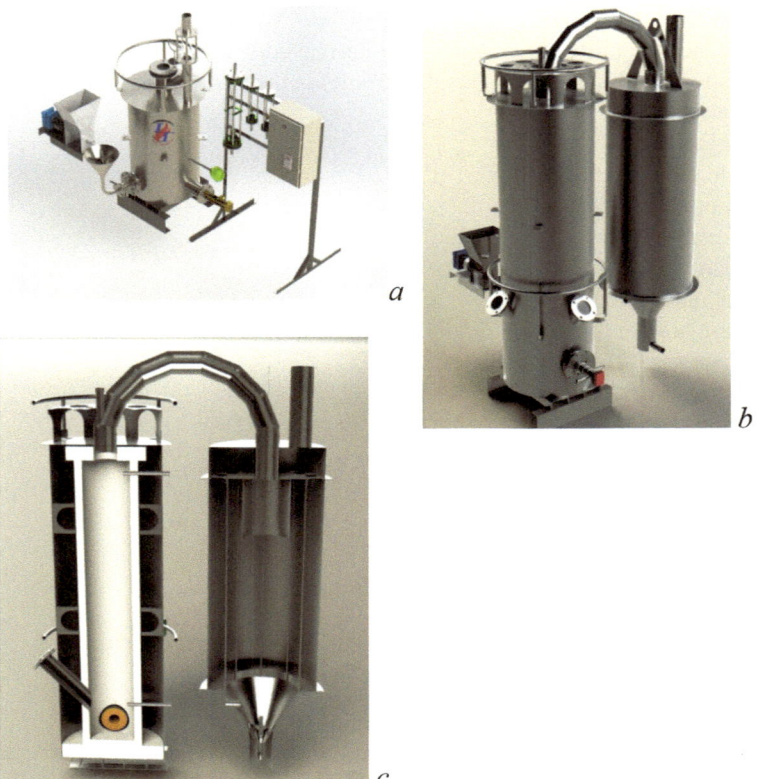

Fig. 2.16 Computer visualization of the new type systems for TEG production. Cyclone-type reactor (**a**) and "opposite-type" reactor (**b**, **c**)

2.6 Unit with Fluidized Bed for Gas–Vapor Activation of Different Carbonaceous Materials for Various Purposes. Design, Computation, and Implementation

We propose the technology of obtaining the promising material with wide specter of application—activated nanostructured carbon. In terms of technical indicators, it will stand next to the materials produced by complex regulations with the use of costly chemical operations. It can be used for such needs: as a sorbent for hemosorption, enterosorption, for creation the newest source of electric current (lithium and zinc air batteries, supercapacitors), for processes of short-cycle adsorption gas separation, etc.

In the study, author gives recommendations concerning design of the apparatus with fluidized bed and examples of calculation of a specific devices. On the whole given information can be used as guidelines for design of energy effective aggregates. Calculation and design of the reactor was carried out using modern software complexes (ANSYS and SolidWorks).

Activated carbon is a porous material produced from different carbonaceous materials: charcoal, coal and petroleum coke, coconut shell, etc. [10]. The essence of activation consists in the pores opening that are in the closed state in the carbon material. It is executed by thermo-chemically method (material is impregnated with a solution of zinc chloride, potassium carbonate, or some other compounds and is heated without access of air) or by treatment with superheated steam or carbon dioxide at a temperature of 800–850 °C. The most widespread activation method is simultaneous supply of incomplete combustion of natural gas and steam in certain proportions in to the activation apparatus. The specific surface of the pores of activated carbon is the most important indicator of its quality, which can reach 600–2200 m^2 per gram, depending on the initial material and activation methods. Thanks to its absorption properties, activated carbon is one of the most common materials which is widely used in various industries: ecology, medicine, civil defense, and military sphere (for example, in the manufacturing of respirators and life-support systems of air-RAID shelters, etc.). Recently, the manufacture of rechargeable batteries of a "new generation" [11] is very important in the production of which activated carbon with high purity and specific surface area is also used.

2.7 The Reactor Type and Its Key Elements. Hydrodynamic Characteristics of Particles of the Processed Material

If the quality and uniformity of the final product is most important factors, it is necessary to apply the apparatus of periodic action. However, this type of apparatus has some drawbacks, for example: the complexity of technological modes of operation and the constructional complexity of loading and unloading elements of the unit. Gas distribution grid of a "cap" type is used to maximize the efficiency of the reactor of fluidized bed. The main function of gas distribution grid is uniform supplying of fluidized gases and generating of the corresponding hydrodynamic mode (depending on the design this element can be manufactured by a porous material or by caps with holes). This key element of the unit is exposed to extreme temperatures. This is in turn places high demands to the used materials of which it is made [12, 13].

The internal volume of activation reactor consists of three main zones: the area under the gas distribution grid, the area above the gas distribution grid (zone of fluidization), and the separation space. The products of incomplete combustion of natural gas are mixed with water vapor in a specific ratio in the first zone of reactor's volume. In the second zone—gases velocity must guarantee initial material fluidizing (initial material at the beginning of the activation process is more dense and larger than at the end of the process). In the third zone—the cross section of the reactor is increased to reduce gases speed which passing through the zone. The gas speed in this zone should provide the return of the main part of small and light material in to the second zone by gravitational forces.

The external view of the reactor of activation and its cross section are shown in Fig. 2.17a–c. At the process of coal activation, combustible gases are emitted which are burnt in the separation zone.

For this purpose, additional burner or air supply is used. This element is not principle, and it will not be considered in this article. After material processing, activated carbon has to be unloaded from the reaction zone and must be cooled without oxygen access. For this purpose, the hermetic cooler has been applied (Fig. 2.18). Usually, the source material is crushed and sieved. Different types of coal have different densities and particle size distribution before and after activation process. For example, the size of the particles prior to activation may be 0.25–10 mm and after the process is less than 0.1 mm. The density of the activated carbon of various types is in the range from 300 to 500 g per liter. In this regard, while the design and calculation of appropriate reactor zones it is necessary to consider these indicators together with the required performance of the unit.

2.8   Calculation Methodology of Material Flows and the Main Parameters of the Unit

The calculation of the material flows and basic dimensions of the reactor (and vice versa) has a certain logic and consistency. Based on the performance of the unit, you must specify the diameter of the reaction zone (diameter of the gas distribution grid). Similarly, it is necessary to determine the "flying speed" of the material based on its type. The example of calculation of activation reactor is given below.

Example of Calculating Flows

- The rate of fluidization of the feedstock with the particulate composition 1–3 mm in the reaction zone $W_f = 0.5$ m/s (the speed of free fall $W_f = 1$ m/s);
- The area of the reaction zone for the apparatus with a capacity of 3 CC per hour $S = 0.785 \times D2 = 0.785$ of $0.212 = 0.0346$ m^2;
- Rate of flue gases (60%) and steam (40%) $G = S \times W = 0.0346 \times 0.5 = 0.0173$ m^3/s (62.28 m^3/h);
- Steam consumption of $G_{steam} = G \times 0.4 = 0.00692$ m^3/s (24.9 m^3/h) or 14.9 L per hour (1 kg water $= 1.67$ m^3 of steam);
- Flow flue gas $G_{f.gas} = G \times 0.6 = 0.01038$ m^3/s (37.368 m^3/h);
- Coefficient of thermal expansion $K = (273 + t)/293 = (273 + 900)/293 = 4$;

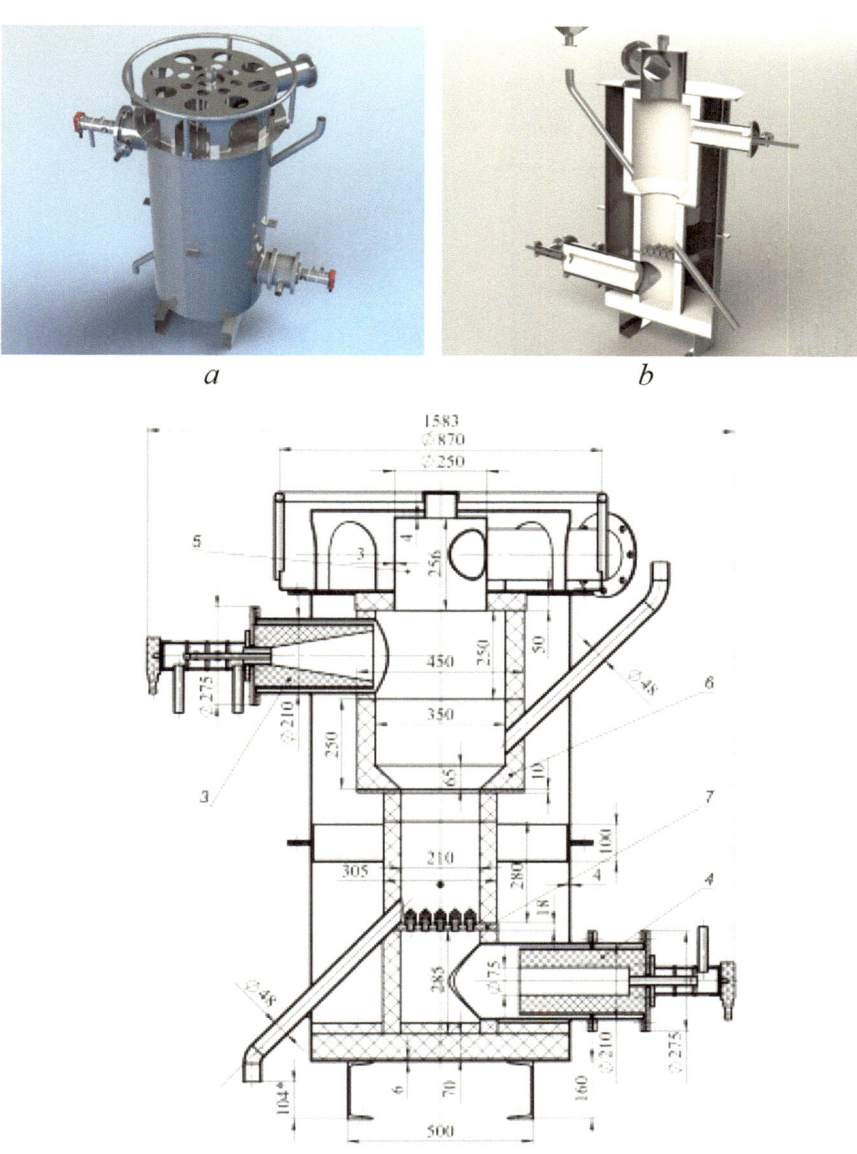

Fig. 2.17 High-temperature coal activation reactor: **a** external view of the reactor, **b** visualization of cross section of reactor, **c** draft of cross section of reactor

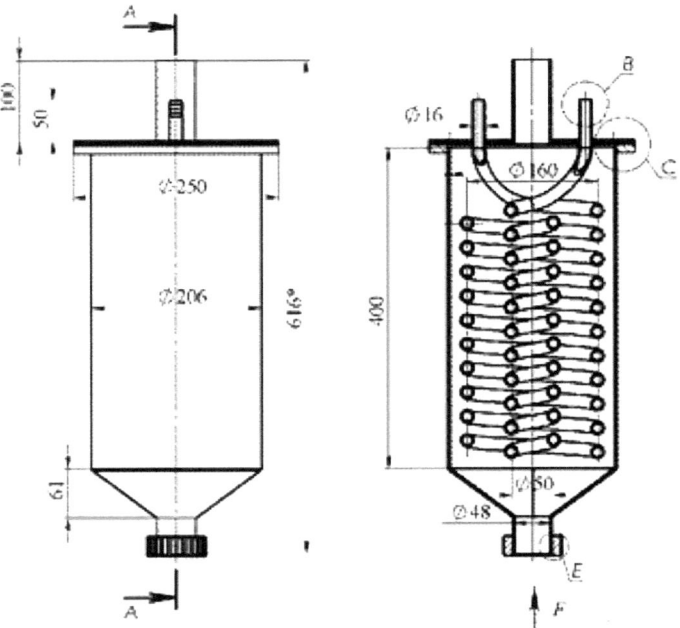

Fig. 2.18 Hermetic cooler for activated coal

- The consumption of the combustible mixture $G_{\text{comb.mix}} = G_{\text{comb.mix}}/4 = 0.002595 \text{ m}^3/\text{s}$.

  Hence, the calculated consumption of fuel and air:

- Gas consumption $G_{\text{gas}} = G_{\text{comb.mix}} \times 0.1 = 0.0002595 \text{ m}^3/\text{s}$ (0.9342 m³/h);
- Air flow $G_{\text{air}} = G_{\text{comb.mix}} \times 0.9 = 0.0023355 \text{ m}^3/\text{s}$ (8.41 m³/h).

For more reliable operation, we apply the gas distribution grid which consists of special caps. The design of the gas distribution grid and appearance are shown in Fig. 2.19.

The number of holes in the cap—6 pieces to ensure the material fluidizing the "living section" of the grid should be 1–3% of the grid area (reaction zone). The calculation of the gas distribution grid is concluded to determining the number of caps and the diameter of the holes in them.

Example of Calculating Hydrodynamic

- Square of grid (reaction zone) $S = 0.785 \cdot D2 = 0.785 \times$ of $0.212 = 0.0346 \text{ m}^2$;
- 2% it is $6.92 \times 10^{-4} \text{ m}^2$ (21 cap);
- The cross-sectional area one cap of 3.29 mm² (6 holes);
- Area of one hole is 5.5×10^{-6} m;

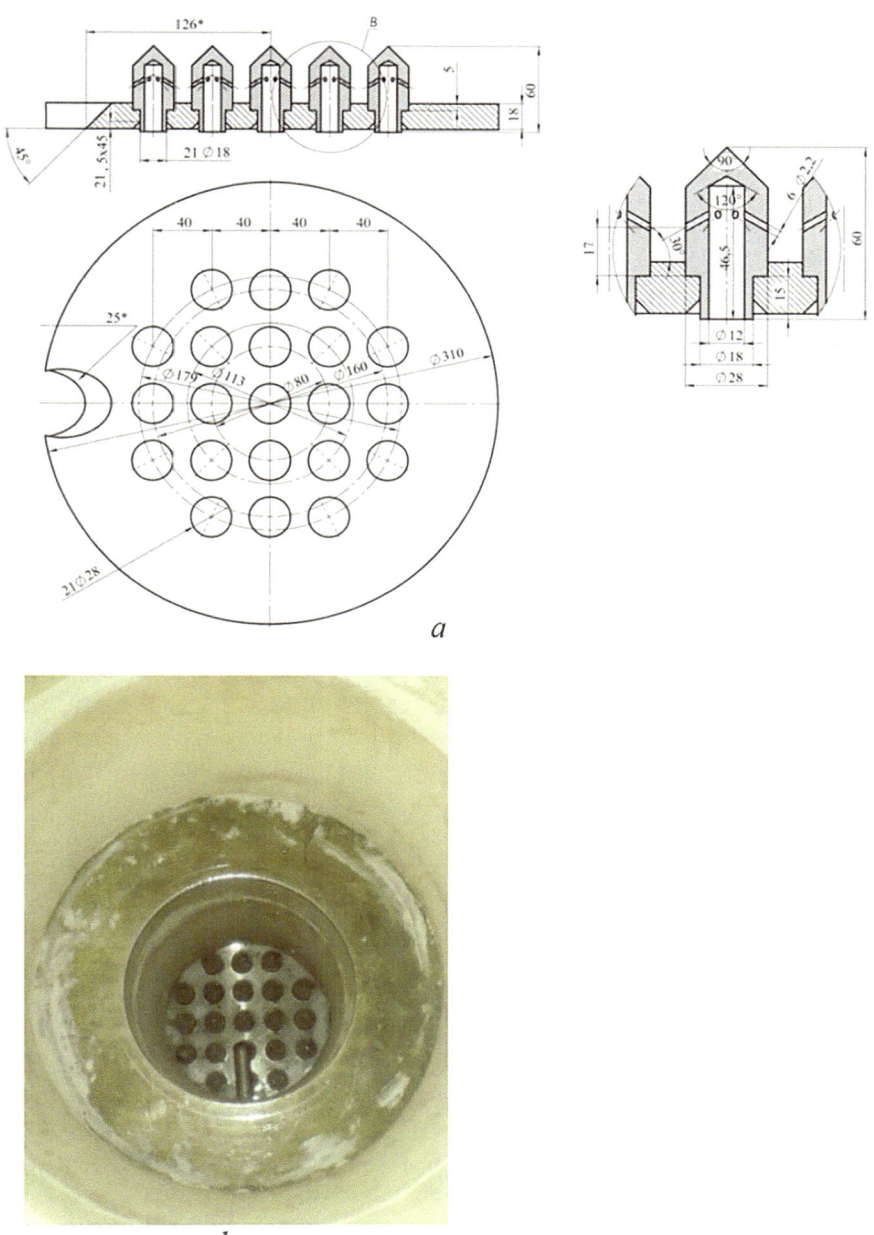

Fig. 2.19 Gas distribution grid: **a** drawing of the gas distribution grid, **b** external view of a gas distribution grid

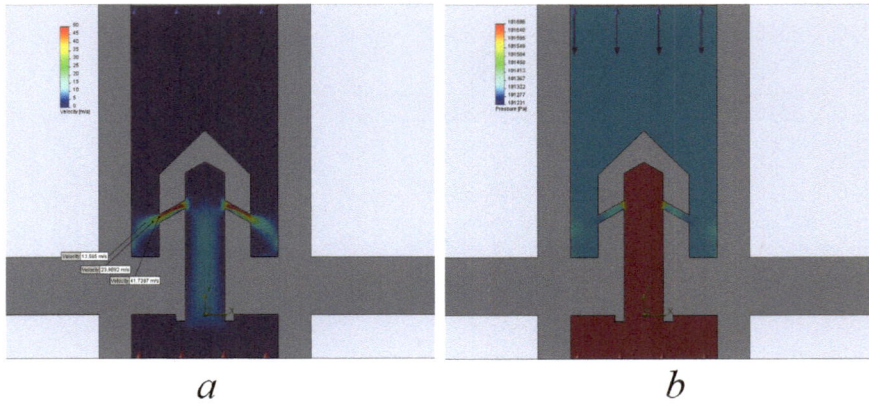

Fig. 2.20 Simulation of gas flow in the holes of the cap of the gas distribution grid: **a** the velocity of gas flow, **b** hydraulic resistance of one cap of gas distribution grid

- The diameter of the hole of the cap $D = (5.5 \times 10^{-6}/0.785)0.5 = 2.64$ mm standard diameter of holes of 2.2 mm.

Thus:

- The gas flow rate single hole cap $G = 0.0173/(21 \times 6) = 1.37 \times 10^{-4}$ m³/s;
- The cross-sectional area of one hole of the cap $S = 0.785 \times 2.2^2 = 3.8 \times 10^{-6}$ m²;
- The calculated average exit speed from the holes of fluidizing agent (flue gas + steam) is $W = G/S = 36$ m/s (the recommended starting speed is 35–40 m/s).

With the help of computer simulation [8, 9], test calculations of the velocities encountered in the hole of the cap are performed, see Fig. 2.20a. Calculation of hydraulic resistance arising from the passage of hot gases through the holes in the cap is shown in Fig. 2.20b.

2.9 CFD-Simulation of Steam-Gas Mixture Flows in the Key Zones of the Activation Reactor

Sustainable fluidization (at the beginning of the activation process) and minimizing of particles removal (at the end of the process) in the appropriate zones of the activation reactor: in the area of fluidizing zone and in the separation zone, are achieved due to calculations and simulations of the processes in these zones [8, 9]. Having data about the volume of material flows in the reactor it is rational to simulate their motion in space of the designing unit to identify "dead" zones and flow turbulence. Figure 2.21a–c shows the velocity and direction of flows, plot of the temperature distribution, and the concentration of water vapor in the separation zone.

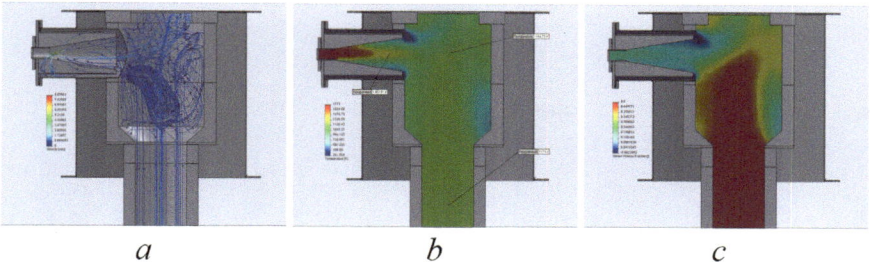

Fig. 2.21 Simulation of gas flow in separating zone of activation reactor: **a** velocity and direction of flows, **b** distribution of temperature, **c** distribution of steam concentration

Burner device is used for generating heat by gas fuel burning. The fuel efficiency is achieved thanks to a good mixing of natural gas and oxidant. The sudden expansion of the channel serves to ignition and stabilization of combustion in the combustion chamber. To make sure of the "quality" vortices, it is possible by simulation [8, 9] of cold gas flows (in case of burning the volume of gases and respectively turbulence are guarantee increased) as shown in Fig. 2.22.

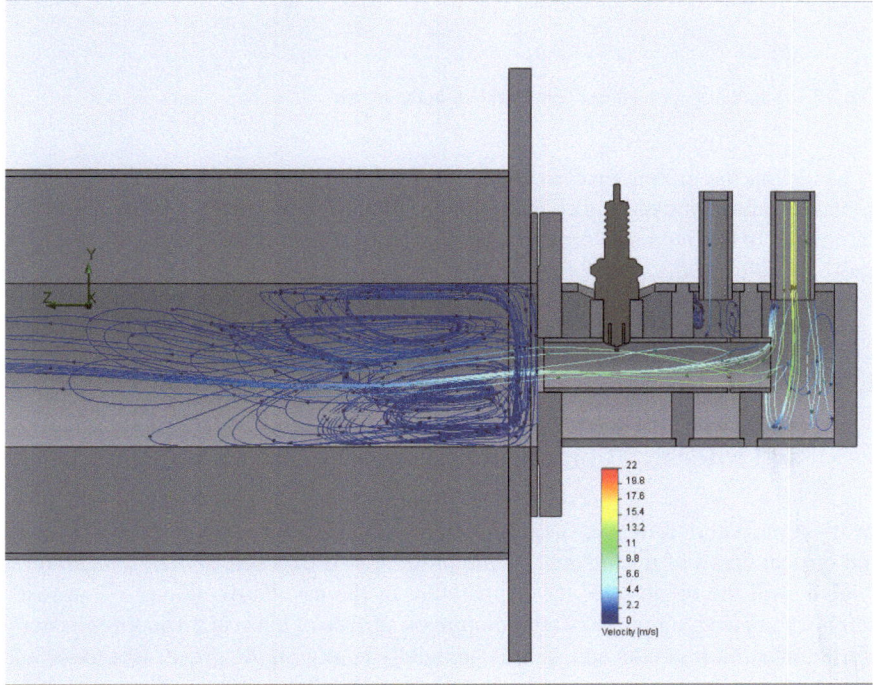

Fig. 2.22 Simulation of flows velocity in the burner

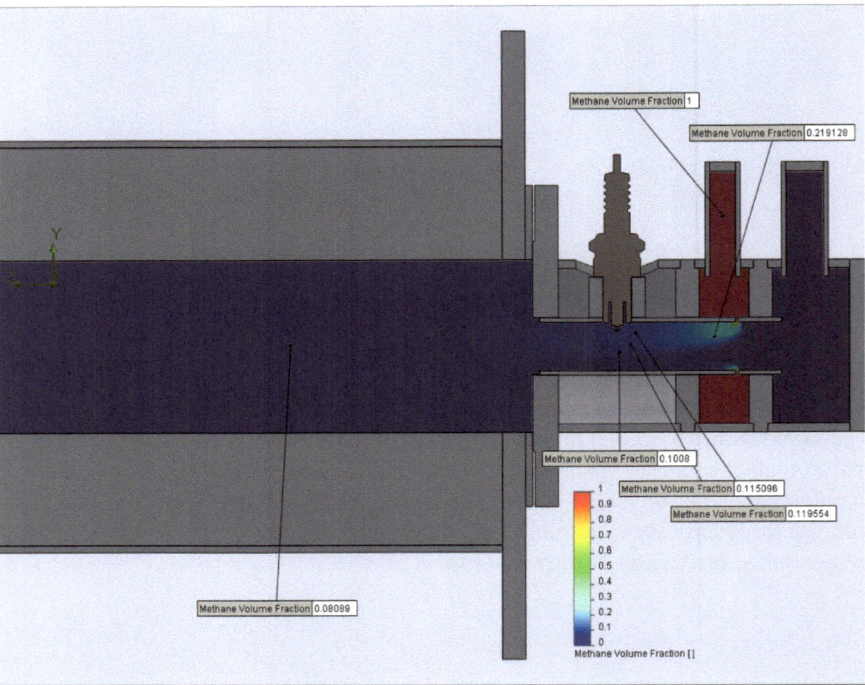

Fig. 2.23 Simulation of methane concentration in the burner

Modeling the mixing process of air and fuel allows you to accurately determine a specific area in the mixing chamber. The concentration of the components will be flammable in this area. Accordingly, in this precisely area it is advisable to place the ignition device of the burner (Fig. 2.23).

2.10 CFD-Simulation of Heat-Gas Mixing of Gas–Vapor Mixture in the Area Under the Gas Distribution Grid in Order to Align the Temperature Field on Its Surface

As mentioned above, the gas distribution grid of the reactor—the most heat-stressed and critical detail of this device. At the process, it is exposed to high temperatures (violation of the mode rules the temperature of the gas distribution grid can reach 1500 °C) and sharp gradient of temperature on surface of the grid. In addition, natural gas combustion products and steam chemically acting on the greed. Hot gases and the cold vapor gases passing through the gas distribution grid in order to avoid its

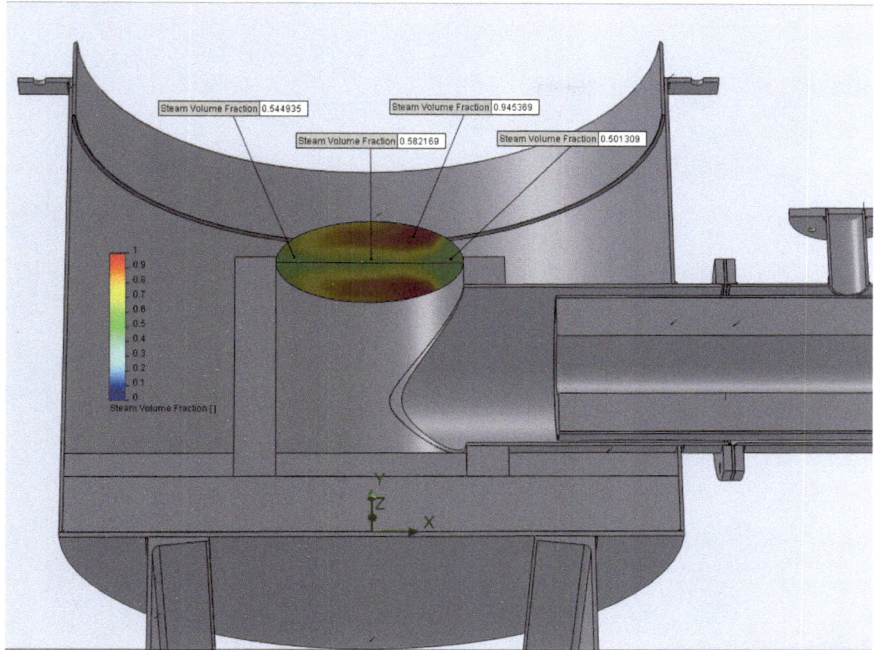

Fig. 2.24 Temperature distributions at the bottom surface of gas distribution grid (variant one of the burner stone construction)

destruction must be thoroughly mixed. Figure 2.24 shows the degree of mixing of gases on the bottom surface of gas distribution grid.

This simulation showed that at the bottom surface of gas distribution grid there is a great temperature gradient. Thanks to the above information, it was decided to change the design of the burner stone (to reduce its length) for better mixing of steam and combustion products. As a result of mentioned design changes on the bottom surface of the gas distribution grid, we managed to achieve less temperature gradient due to better mixing of the flows (Fig. 2.25).

The direction and velocity of the flow in the mixing chamber of vapor and combustion gases are shown in Fig. 2.26.

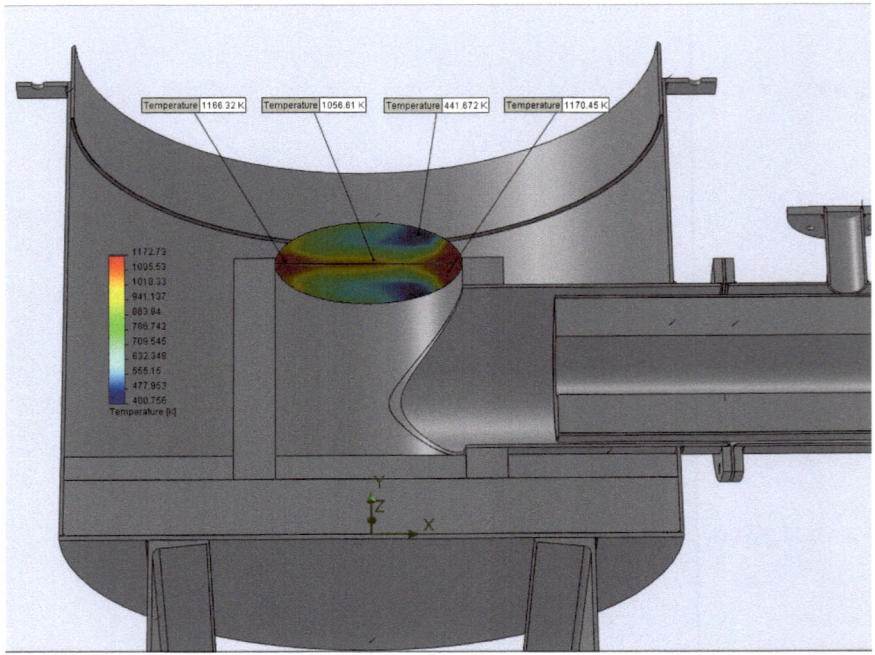

Fig. 2.25 Simulation of temperature distribution at the bottom of gas distribution grid (variant two of the burner stone construction)

3D designing of the reactor and its individual elements combined with the modeling of physical processes occurring in the reactor manage to avoid the conceptual errors at the initial stage of manufacture of the device. Such errors would result to rapid failure of the whole unit and therefore—to great financial losses for the enterprise.

2.11 Design of the Unit in 3D (SolidWorks). Strength Calculation of Construction Elements of Activation Unit (ANSYS)

Knowing the mass of the reactor, it is possible to design a carrier frame and a platform for maintenance of unit's elements and auxiliary equipment (Fig. 2.27).

The calculations test [9] of the supporting frame on a strength showed a peak stress in the material of 31.3 MPa, which gives the opportunity to use cheaper steel or to reduce metal consumption of the design (the yield strength for cheap carbon steel is 220 MPa), see Fig. 2.28.

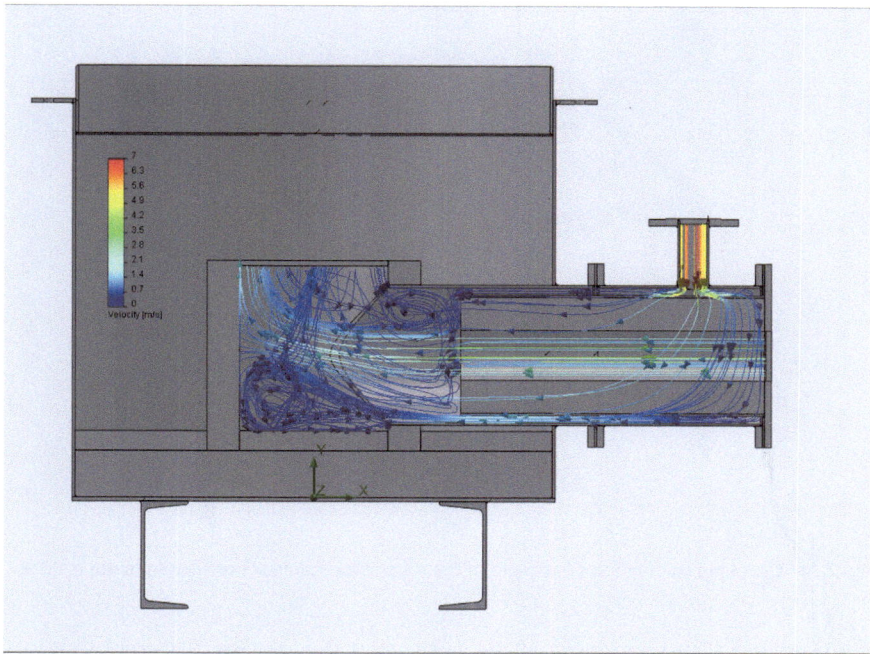

Fig. 2.26   Direction and velocity simulation of combustion products flows and steam in the mixing chamber

Fig. 2.27   Visualization of frame for services of activation reactor

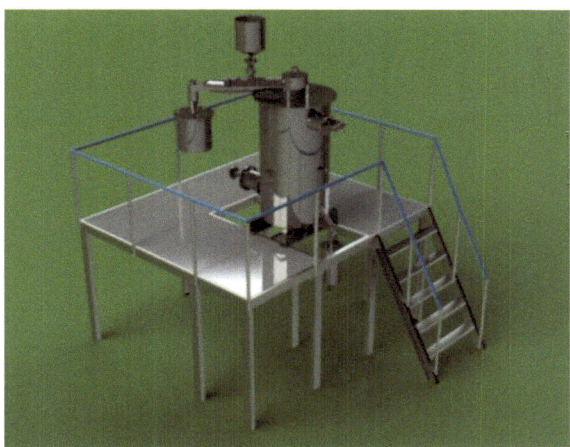

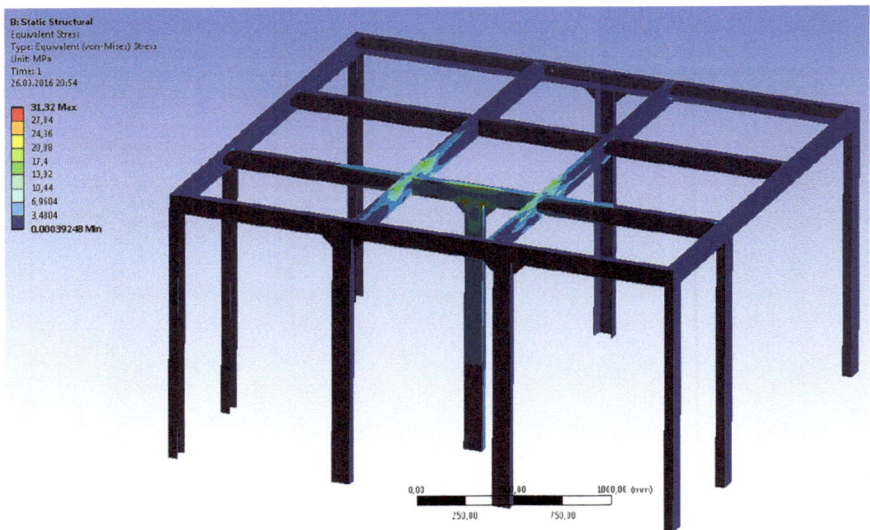

Fig. 2.28 Checking calculation of strength of the supporting frame for equipment maintenance

Running and Commissioning of the Activation Reactor for Different Types of Processing Materials. Preparing of the unit to work is concluded in it connection to the mains, water supply, air, and gas lines as well as verification of the steam generator and other accessory before you turn them on. Next, the reactor will start and running to nominal mode of air, gas, and steam consumption. Then, material is supplied to activation. Figure 2.29 shows a commissioning moment of the unit.

The first runs of the reactor were made without the protective cover. Thanks to this we managed to fix by photo and video of the heated gas distribution grid without coal and directly the activation process (Fig. 2.30a, b).

Technical Characteristics of the Activation Unit

- Productivity on the finished product—1–3 kg/h (depending on the type of row materials);
- The natural gas consumption of 1 m^3/h;
- Steam consumption—10–15 l/h;
- Working temperature of the activation process at 900 °C;
- Density of activated carbon 0.42–0.5 g/sm^2;
- The specific surface of the coal 1300–2000 m^2/g (depending on the raw material and activation mode);
- Device is of periodic action (duration of one cycle 1–2 h).

Fig. 2.29 Setup and tuning of equipment: **a** setup of gas and air consumption, **b** tuning of burner automatic system

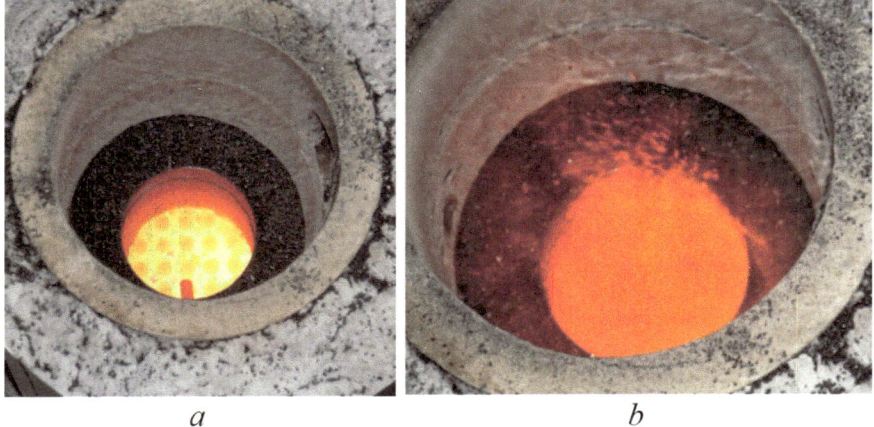

Fig. 2.30 Gas distribution grid in the processes: **a** hot distribution grid, **b** fluidized bed of activated carbon

Summary

The Gas Institute of N.A.S. of Ukraine is a leading scientific institution that has many years of experience in the development and implementation of technologies and technological equipment for the production of activated carbon on an industrial level. Modern and energy saving units of the new generation are the result of many years of experience and CFD technologies. For example, unit wich implementation carried out in the USA is (Argonne National Laboratory) in 2013 is described in this article. The Gas Institute of NAS of Ukraine is constantly working on even more sophisticated, resource-saving technologies, and equipment to produce high quality activated carbon from different raw materials.

References

1. Yaroshenko AP, Savoskin MV (1995) High quality intumescent graphite intercalation compounds for new approaches to chemistry and technology. J Appl Chem 68(6):1302
2. Chernish SG, Karpov II, Prihodko GP, Shay VM (1990) Physicochemical properties of graphite and its compounds, vol 15. Naukova Dumka, Kiev, p 200
3. Ionov G, Kuvshinnikov V (2003) Physical and chemical properties of low-density carbon materials. In: Carbon: fundamental problems of science, material science, technology: proceedings of the 2nd international conference, vol 111, Moscow, 15–17 Oct 2003, pp 42–56
4. Avdeev VV, Nicholas IV, Monyakina LA (2008) Flexible graphite foil and process for its preparation, vol 56. Machinery, Moscow, p 423
5. Barsukov I (2000) Applications for battery carbons. Battery Power Prod Tech 4(9):30
6. Dmitriev VM, Kozhan AP, Ryabchuk VS, Strativnov EV, Bondarenko OB (2012) Pat. 104098 Ukr., IGC V09S1/00. A method of purifying water and soil from oil and oil graphite sorbent. Publ. 12.08.2013
7. Strativnov EV, Kozhan AP, Bondarenko BI (2011) Patent. 99875 Ukr., MPK C01B 31/04. Method for producing expanded graphite. Published 10 Oct 2012
8. Chigarev AV, Kravchuk AS, Smalyuk AS (2004) ANSYS for engineers. The reference guide, vol 5. Machinery, Moscow, p 511
9. Alyamovsky AA (2005) Engineering calculations in SolidWorks, vol 12. DMK Press, St. Petersburg, p 815
10. Benaddi H, Bandosz TJ, Jagiello J, Schwarz JA, Rouzaud JN, Legrasc D, Beguin F (2000) Surface functionality and porosity of activated carbons obtained from chemical activation of wood. Carbon 38:669–674
11. Peng C, Zhang SW, Jewell D, Chen GZ (2008) Carbon nanotube and conducting polymer composites for supercapacitors. Prog Nat Sci 18(7):777–788
12. Boothroyd RG (1975) Flowing gas-solids suspensions. Translated from English by Danilin SV, Spokoyniy FE. Publishing House "Mir", Moscow, 373 p
13. Sinnott RK (2005) Chemical engineering, vol 6. Elsevier Butterworth-Heinemann, 1038 p

Chapter 3
Production Technology and Application of Materials Based on Thermally Expanded Graphite

Abstract One of the most progressive trends in modern science and technology is the development of energy-efficient technologies for synthesis of nanomaterials. Nanolayered graphite (thermally expanded graphite) is one of the key nanomaterials based on carbon. Due to its unique properties (chemical and thermal stability, ability to form without a binder, elastic, etc.), it can be used as an effective absorber of organic substances, the material for seals manufacturing for such important industries as gas transportation, automobile, etc. Thermally expanded graphite is a promising material for hydrogen and nuclear energy industry. The development of thermally expanded graphite production is resisted by high specific energy consumption at its manufacturing and by some technological difficulties. Therefore, the creating of energy-efficient technology for it production is a very actual point.

3.1 Research and Analysis of Dispersion and Cavitation Processes

The term "graphene" for an individual graphite layer was introduced in 1994 and refers to intercalated graphite intercalation compounds. It is proposed by analogy with the names of polycyclic aromatic hydrocarbons (anthracene, coronene, etc.). Currently, studies in the field of graphene are not limited to single-layer samples; structures containing two or more (up to 100 pieces) graphene layers are also of interest [1–5].

Composite materials based on graphene are used in energy storage devices (as electrodes in batteries and capacitors) [6]. The electrodes produced in this way have high conductivity and a large surface area, which is necessary to achieve ultra-high electrical characteristics of batteries [7]. When optimally mixed with the base, it forms a three-dimensional electron-conducting network of cathode material. This increases the conductivity of the material and the charging speed of the battery. Given the advantage of electronically conductive graphene additive over other types of carbon, it can be expected to be widely used in commercially available high-performance lithium-ion batteries.

Methods for obtaining graphene can be divided into four groups: (1) mechanical separation of layers [2–5], (2) chemical deposition on a substrate from the gas phase [8], (3) organic synthesis [9], and (4) a chemical method using colloidal dispersions on based on compounds containing graphene layers [10]. In this paper, it is proposed to consider several mechanical methods for obtaining multilayer graphene: using cavitation [11–13], ultrasound, and due to the electrodynamic explosion of the "Yutkin effect" [14]. All of them consist of a sequential two-stage mechanical action on thermally expanded graphite (TEG) in a certain way [15]. Graphene has been made by different methods. The first was chemical vapor deposition (CVD) and epitaxial growth, such as the decomposition of ethylene on nickel surfaces. These early efforts (which started in 1970) were followed by a large body of work by the surface-science community on "monolayer graphite". The second was the micromechanical exfoliation of graphite. This approach, which is also known as the "Scotch tape" or peel-off method, followed from earlier work on micromechanical exfoliation from patterned graphite. The third method was epitaxial growth on electrically insulating surfaces such as SiC and the fourth was the creation of colloidal suspensions.

- Mechanical method: The cheapest, most technologically advanced and relatively environmentally friendly method for obtaining graphene. Depending on the selected technological process, it is possible to vary some properties of the final product. This makes it possible to adjust to the specific required parameters for graphene.
- Chemical vapor deposition onto a substrate: This method makes it possible to obtain the highest quality large-area graphene monolayer. Currently, this method is used exclusively for the purpose of studying the properties of graphene. The method is quite expensive and not technologically advanced.
- Chemical method using colloidal dispersions based on compounds containing graphene layers. Graphene obtained by this method contains residues of chemical compounds and has an unpredictable shape and number of layers. These properties strongly depend on the initial graphite. This method is not environmentally friendly and is used where there are no strict quality requirements for graphene. If graphene or very thin platelets of multilayer graphene could be produced on a large scale by CVD from various precursors, new routes for creating colloidal suspensions might also be found, and the supply of graphene/few-layer graphene might be enormously increased.

It should be noted that non-graphene carbon additives prevent the dense stacking of graphene structures, improving the cathodic electrochemical characteristics [16]. Graphene and its derivatives are widely used in Li–S batteries, effectively improving their electrochemical characteristics due to outstanding mechanical strength, exceptional conductivity, and large specific surface area. Graphene sheets interact with lithium polysulfide and thus improve the overall use of the functional material. Graphene materials have also been used as a conductivity enhancer in solid electrolytes.

This is a conductive additive for battery matrixes. Leading producers of battery-grade expanded graphite are Japanese Nippon Carbon and Swiss Timcal, Ltd. Their products were developed exclusively for alkaline batteries and need significant modification in terms of particle size and surface chemistry before application in non-aqueous systems. To keep pace with the emerging need for high-voltage cathodes, some new and improved forms of expanded graphite are needed (Few layers Graphene). This is what the experts at the Gas Institute of N.A.S. of Ukraine in Kyiv focused on as part of their development effort.

The proposal is based on the task of improving the method of obtaining graphene from natural graphite, in which as a result of high-speed heating of oxidized graphite to obtain thermally expanded graphite (TEG), before dispersion and reduction of oxidized graphite, subsequent mixing of thermally expanded graphite with water to obtain water cavitator greatly simplifies the technology of graphene. Due to this, the specific energy and material costs for the process are reduced, which allows to use the proposed method for the production of graphene on an industrial scale. The proposed method is carried out as follows. The natural graphite powder is oxidized with concentrated sulfuric acid, followed by treatment with a mixture of $KMnO_4$ and H_2O_2, then the oxidized graphite is subjected to hightemperature heating (thermal shock) to a temperature of 800–900 °C, resulting in thermoexpanded graphite, which is mixed with water in a ratio of 1:3, the resulting suspension is subjected to pre-dispersion (compaction) in a mechanical cavitator, and ultrasonic dispersion with a power of 1200 W at a frequency of 30 kHz. The proposed method significantly simplifies the technology of graphene production, which reduces the specific consumption of energy and expensive materials in its production, which, in turn, makes it possible to use the proposed method for the production of graphene on an industrial scale. The resulting suspension is subjected to pre-dispersion (compaction) in a mechanical cavitator, and then—the final dispersion by ultrasound with a power of 1200 W at a frequency of 30 kHz.

3.2 Research and Analysis of Dispersion and Cavitation Processes. CFD Modeling Using ANSYS Software

By its nature, thermally expanded exhibits hydrophobic properties, which is a negative side for its further treatment in aqueous suspension. Therefore, the first step of the task was to make the "pseudohydrophilic" thermoexpanded graphite.

The cavitation method is suitable for this purpose. Cavitation was performed on two types of equipment: a centrifugal dispersant and a Venturi tube (Figs. 3.1 and 3.2). The cavitation centrifugal dispersant performs the function of mixing TEG with water, its globulation to obtain a suspension suitable for further processing (Fig. 3.3).

The impeller has a specific geometric shape, thanks to which the material is efficiently processed (Fig. 3.4). TEG is very poorly wetted with water—it is a hydrophobic material. However, due to its treatment in a high-speed centrifugal

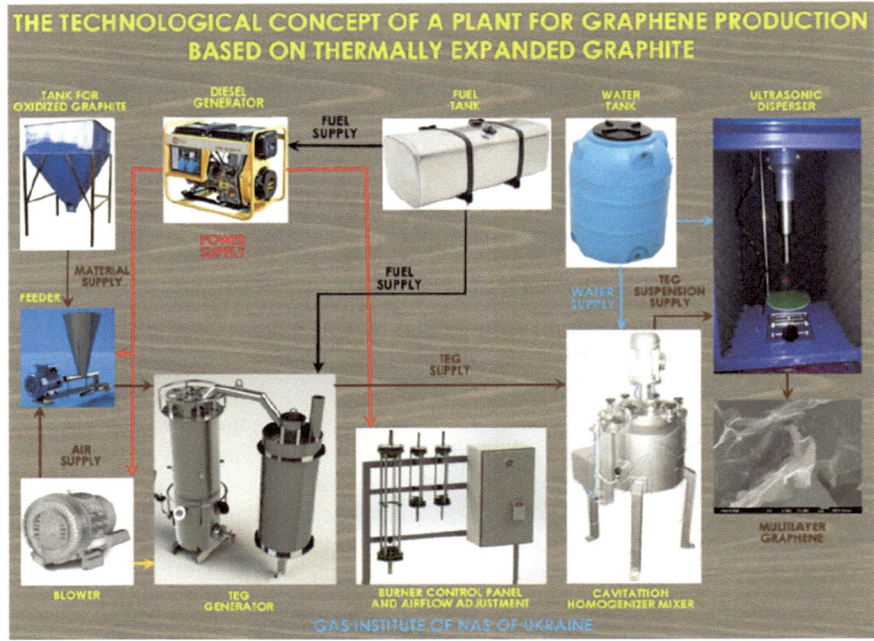

Fig. 3.1 Technological scheme of graphene production from thermally expanded graphite (TEG)

Fig. 3.2 Cavitation centrifugal dispersant

cavitation dispersant, it temporarily acquires hydrophilic properties (which we called "pseudohydrophilic" properties). A mixture of water (3 l) and TEG (1 l) was treated in a cavitation-type dispersant for 5.10, 30 min. Then, the samples were dried on filter paper.

As can be seen from Fig. 3.5, the degree of processing of TEG in the cavitation apparatus has almost no effect on the size of the globules (approximately 0.2 mm) and its pseudohydrophilicity—the suspension peels off after about a day. Therefore: 1—the next technological operation must be carried out as soon as possible after

Fig. 3.3 Cavitation disperser impeller

Fig. 3.4 Pseudohydrophilic properties of TEG

dispersion; 2—it makes no sense to spend a lot of time processing the TEG. Experiments have shown that it is sufficient to wait for stable mixing of TEG with water to obtain a homogeneous suspension. It takes about 1 min (Fig. 3.6).

Photomicrographs of the stratified suspension of TEG show an identical structure of the material that settled and floated after the day of settling, which in turn proves the sufficient quality and time of its processing. The reason for this stratification is the presence in the processed material of micro-air bubbles, which pulls them to the mountain (Fig. 3.7).

Being in the environment of water (suspension), TEG is ready for the next technological stage—processing in an ultrasonic dispersant. Where, due to ultrasonic vibrations of the medium (water), which is almost not compressed—separate layers of graphene are detached from the multilayer "bundles" of TEG (Fig. 3.8).

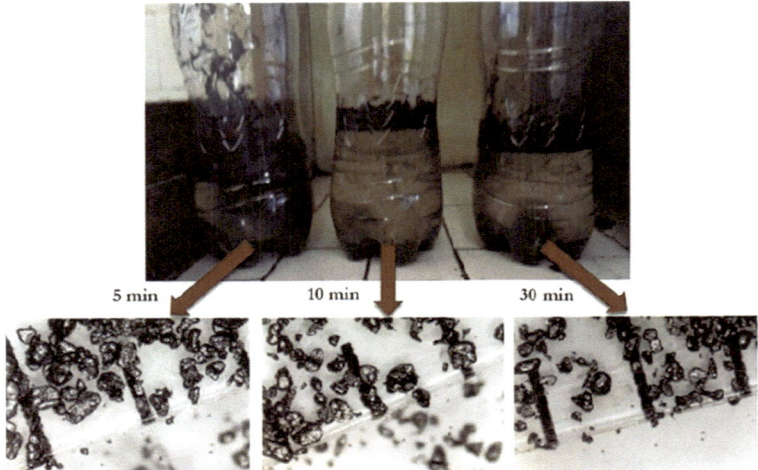

Fig. 3.5 Aqueous suspension of TEG (top). Processing time in the cavitation device: 5, 10, and 15 min. Globules of TEG (below)

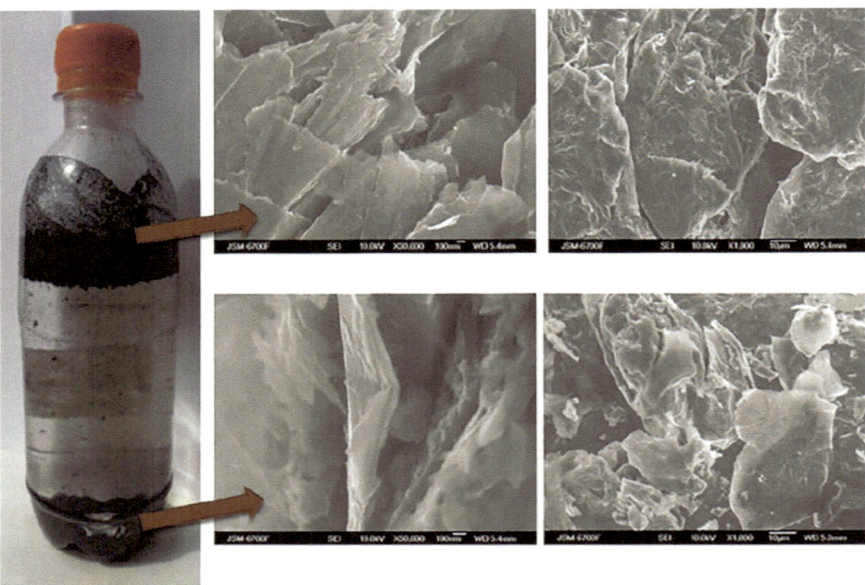

Fig. 3.6 Sedimentation (1 day) and layering of TEG after processing in the cavitator

Fig. 3.7 Prepared (pseudohydrophilic) TEG

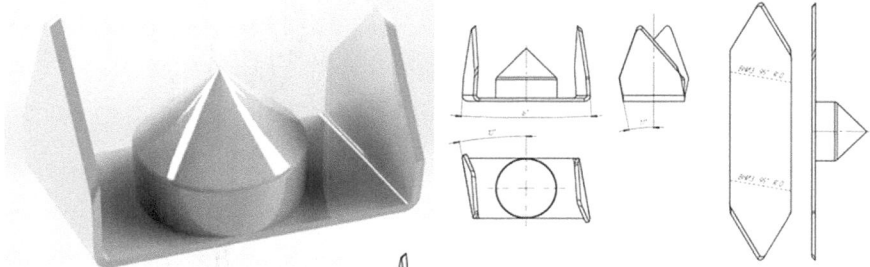

Fig. 3.8 3D model of the impeller

The impeller has a specific shape of its blades, thanks to which effective cavitation and mixing of the suspension is carried out (Fig. 3.9).

The effect of cavitation occurs directly near the blades of the cavitator. Therefore, the density of the calculated grid is much higher near the walls—much higher than in the rest of the volume.

From Fig. 3.10, it can be seen that the absolute flow velocity at the sharp edges of the blades is approximately $V = 6$ m/s, which in absolute values is not a large figure. That is, we can say about the efficiency of the unit, which nevertheless provides the desired effect—cavitation (Fig. 3.11). The percentages of vortex viscosity are clearly distributed over the studied areas: in the total volume, in the immediate vicinity of the impeller blades, and on the surface of the impeller. In water, vortex motion occurs in those parts of the flow where the force of viscosity is manifested to the greatest extent—in the layer near the surface of the streamlined body, in the so-called boundary layer filled with strongly turbulent medium (Fig. 3.12).

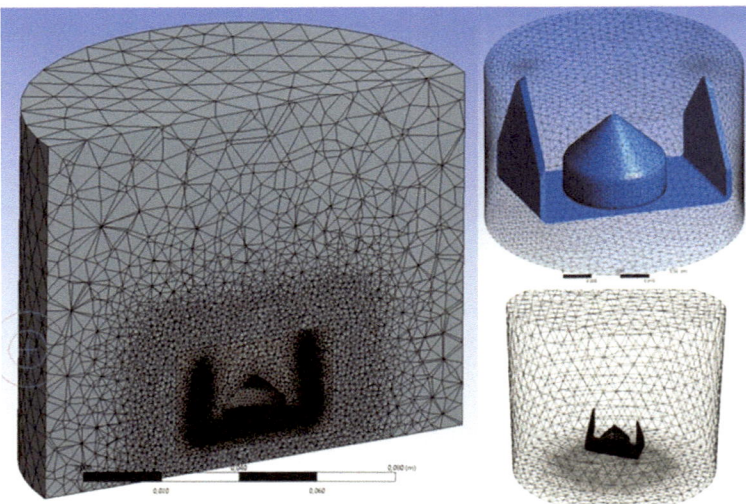

Fig. 3.9 3D model of the impeller by the "ANSYS" modelling

Fig. 3.10 Direction and
speed of flows

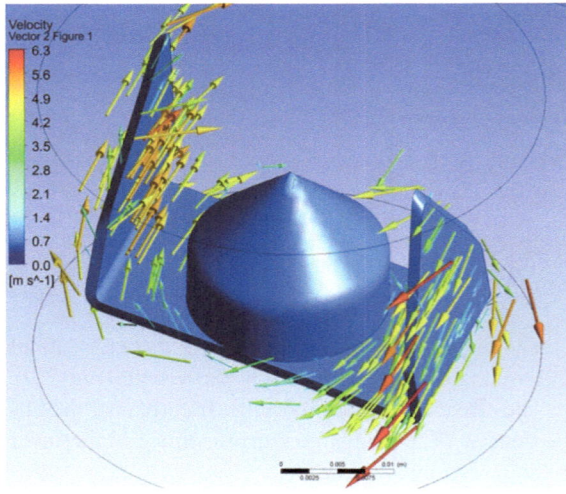

These vortices receive energy from the middle stream as well as from each other. So, these are energy vortices that contain most of the energy. They have large fluctuations in flow rate and have a low frequency (Fig. 3.13).

Turbulence—three-dimensional non-stationary fluid motion, in which the stretching of vortices creates a continuous distribution of chaotic pulsations of flow parameters (velocity, pressure, etc.) in the range of wavelengths from the minimum determined by viscous forces to the maximum determined by the boundary flow conditions.

Fig. 3.11 Eddy viscosity
(vortex viscosity), [Pa s]

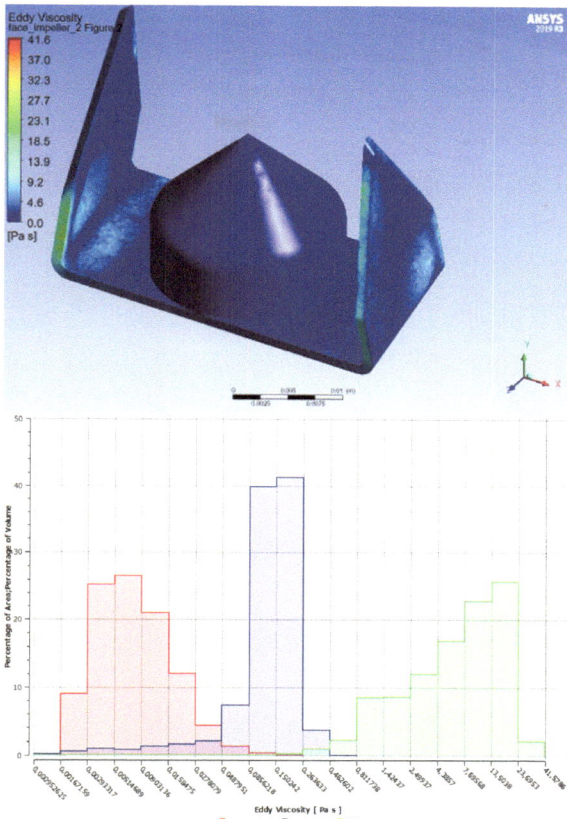

The analysis showed that the chosen geometry of the impeller blades, its speed, and angle of attack provide the necessary effect—cavitation and the necessary mixing throughout the volume (Fig. 3.14).

White color on the model of the impeller-gas bubbles formed by cavitation. It is quite natural that cavitation (formation of cavities inside the liquid filled with gas, steam, or their mixture, i.e., violation of the continuity of the liquid) occurs where the most stressful conditions occur—high turbulence, relative flow velocities, and more. Thus, near the streamlined body (impeller blade) is created quite clearly limited "cavitation zone", filled with moving bubbles. The contraction of the cavitation bubble occurs at high speed and is accompanied by a hydraulic shock, the stronger the less gas the bubble contains. Due to the high destructive force of cavitation and the instantaneous propagation of this force throughout the volume of the suspension— the treated material is destroyed and stratified. As a result, we get multilayer globular graphene.

The phenomenon of cavitation can be achieved in different ways: for example, using a Venturi tube (Figs. 3.15, 3.16, and 3.17).

Fig. 3.12 Eddy dissipation
[m^2/s^3]

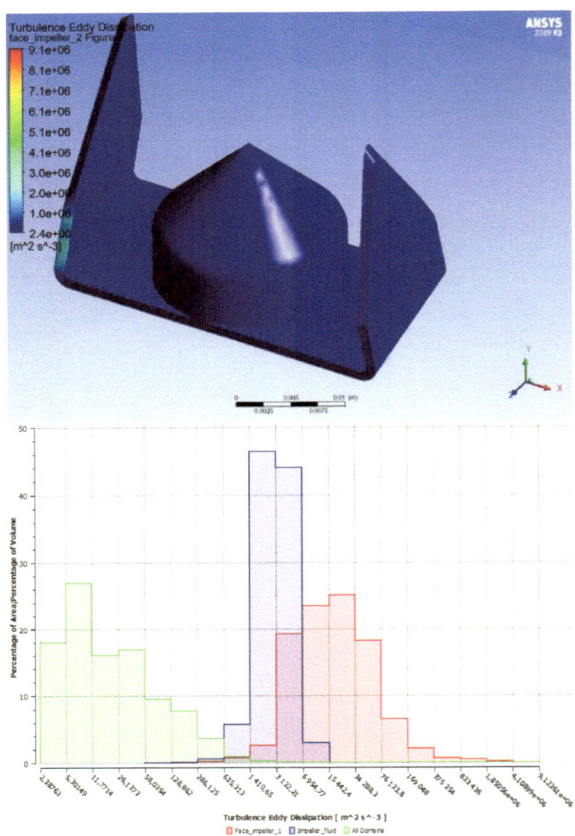

The red color on the diagram indicates the cavitation zone—the lower diagram.

On the upper diagram at the outlet of the Venturi tube, the flow velocity reaches 36 m/s—red (Fig. 3.18).

The container with the water-graphite suspension is subject to ultrasonic treatment. Figure 3.19 shows a scheme consisting of the so-called horn of the ultrasonic dispersant, a liquid medium consisting of liquid (in our case—water) and pseudohydrophilic prepared TEG, as well as a metal vessel. Each of the three participants in this mechanical system is rigidly connected to each other and has its own mass and shape, and therefore its own resonant frequency (Fig. 3.20).

Apart from other participants in the system, the "horn" of the dispersant has a resonant frequency of about 21,000 Hz (Fig. 3.21).

The system, consisting of "horn" and liquid, has a resonant frequency—26,750 Hz. In this case, the mass of material in the liquid can be neglected (Fig. 3.22).

The mechanical system "horn", liquid, and reactor vessel has a negative peak in addition to resonance. This is due to the phenomena of interference and different speeds of sound in different environments (speed of sound in metal 5000 m/s).

Fig. 3.13 Turbulent kinetic energy, $[m^2/s^2]$

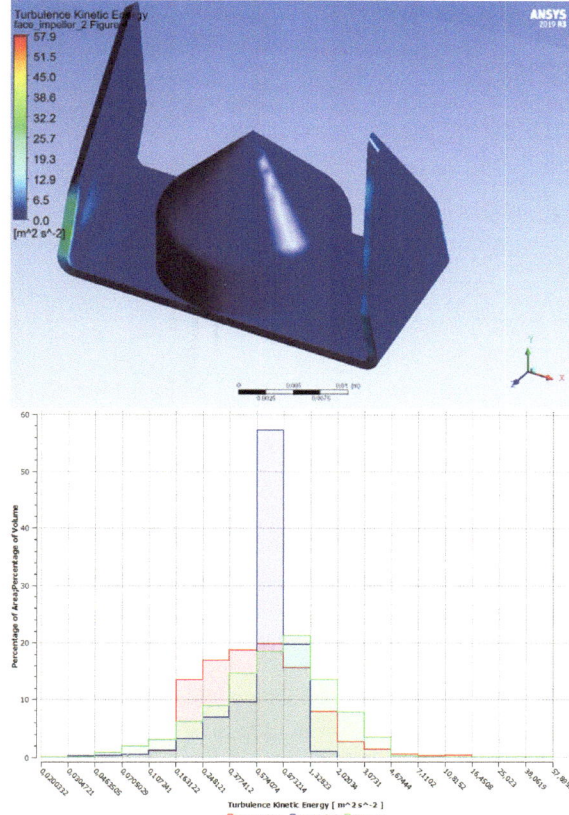

Fig. 3.14 Solid model impeller with cavitation

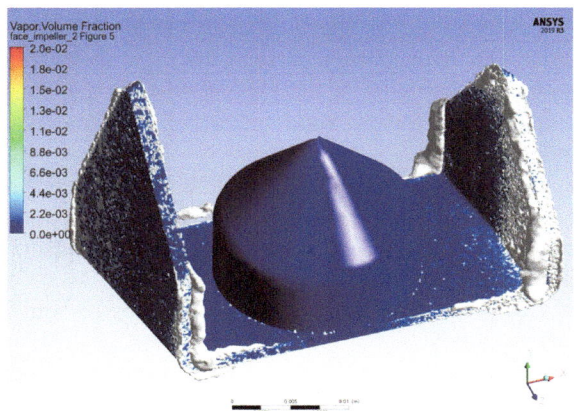

Fig. 3.15 Cavitation with a
Venturi tube

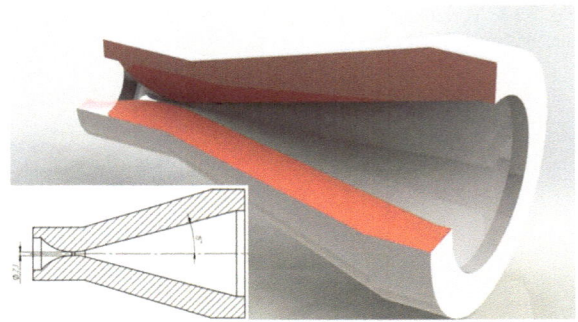

The appearance of the material prepared for processing and after its processing eloquently demonstrates the effectiveness of ultrasonic processing. Figures 3.23 and 3.24 show macro and microphotographs of the treated material, respectively. There are clearly visible individual graphene layers (Fig. 3.25).

Another device that causes the destructive phenomenon of cavitation is an electric discharge device that works on the principle of gradual accumulation of electrical energy and instantaneous discharge. The electrodes of this device are immersed in the liquid. In the case of an instantaneous discharge, all the electrical energy accumulated in the capacitors is released, causing an underwater explosion (hydrodynamic impact). This phenomenon is called the "Yutkin effect".

The electrohydraulic effect is a high-voltage electric discharge in a liquid medium. During the formation of an electric discharge in a liquid, energy is released over a relatively short period of time. A powerful high-voltage electric pulse with a steep leading edge causes various physical phenomena.

In particular—the emergence of ultra-high pulsed hydraulic pressures, electromagnetic radiation in a wide range of frequencies—sometimes to X-ray, cavitation. Electrohydraulic discharge occurs when a pulse voltage of sufficient amplitude and duration is applied to the fluid, resulting in the development of an electrical breakdown. The duration of the leading edge of the discharge current pulse, which is a prerequisite for the Yutkin effect, is from fractions of a microsecond to several microseconds.

In the picture, you can see a container with a water-graphite suspension through which an electric arc passes. Due to the high-voltage electric pulse, there is a stratification of TEG with the formation of graphene bundles, 7–10 layers in each (Fig. 3.26).

Due to the fact that water is not a compressed liquid, micro-explosions grind TEG throughout the tank (Figs. 3.27 and 3.28).

An underwater explosion simulation was performed; the plots show the pressure that develops in the liquid when an electric pulse is applied.

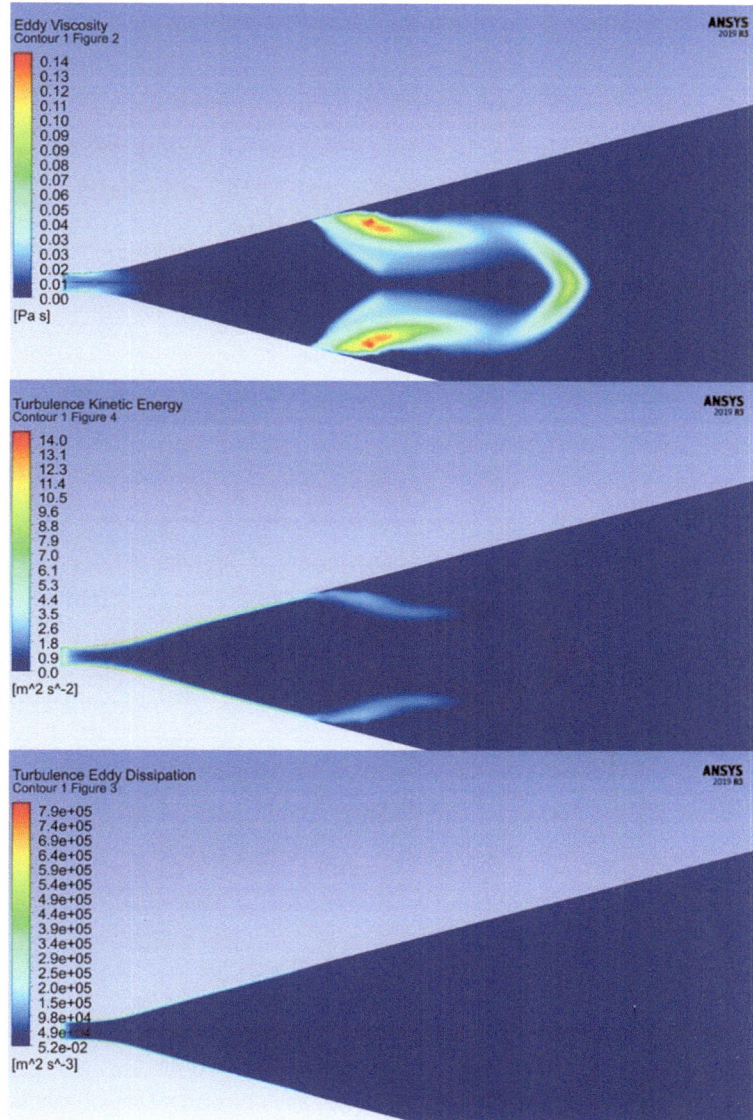

Fig. 3.16 Plots of vortex viscosity, kinetic energy, and vortex scattering

Summary

The problem of obtaining multilayer graphene using thermoexpanded graphite as a raw material was successfully solved. The technological chain consisted of two main stages: obtaining a water-graphite suspension by cavitation and grinding graphite

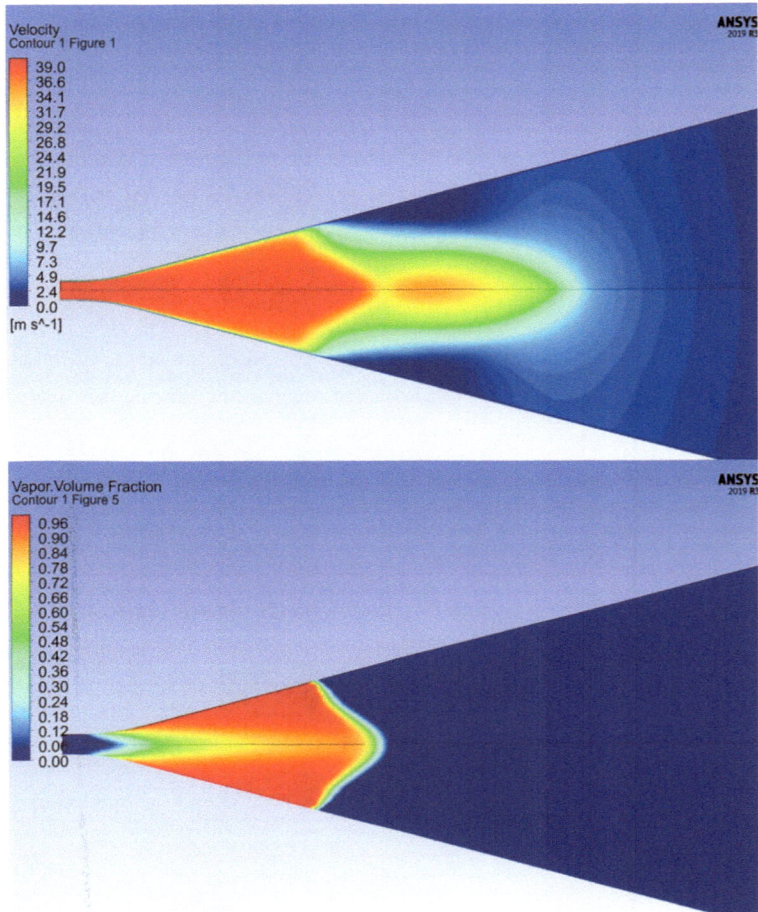

Fig. 3.17 Flow rate and component of gas fraction (cavitation)

globules on graphene plates using ultrasonic or electrohydraulic impact by Yutkin's method. Thanks to CFD modeling, the optimal geometry of the impeller, the speed of its rotation to obtain a cavitation phenomenon, and uniform mixing throughout the volume were determined. Experiments have shown the inexpediency of long-term sonication and limited to a time of 5 min, during which the necessary grinding. The result of the experiments—multilayer graphene (Few-Layer Graphene), which is a pack of 10–20 graphene plates in each.

Fig. 3.18 Ultrasonic
dispersion

Fig. 3.19 Grid and
calculation areas

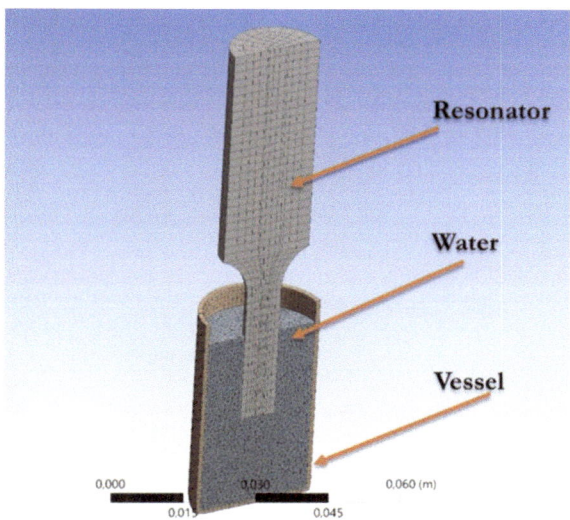

3.3   Graphite-Graphene Composite as an Anode
for Lithium-Ion Batteries

Graphite-graphene composites (GGC) have been obtained as a result of mechanical
treatment of thermoexpanded graphite (TEG). Raman spectroscopy proves the presence of ordered graphene in the GGC. The predominant formation of no more than
five graphene sheets in the material is concluded from Raman data and SEM micrographs. Electrochemical tests of GGC samples show that in spite of quite low specific
discharge capacity (290 mAhg^{-1}), a 20-fold increase in current density (from 50 to
1000 mAg^{-1}) does not lead to a change in the specific capacity upon deintercalation of lithium ions. This feature favorably differs the material studied from existing

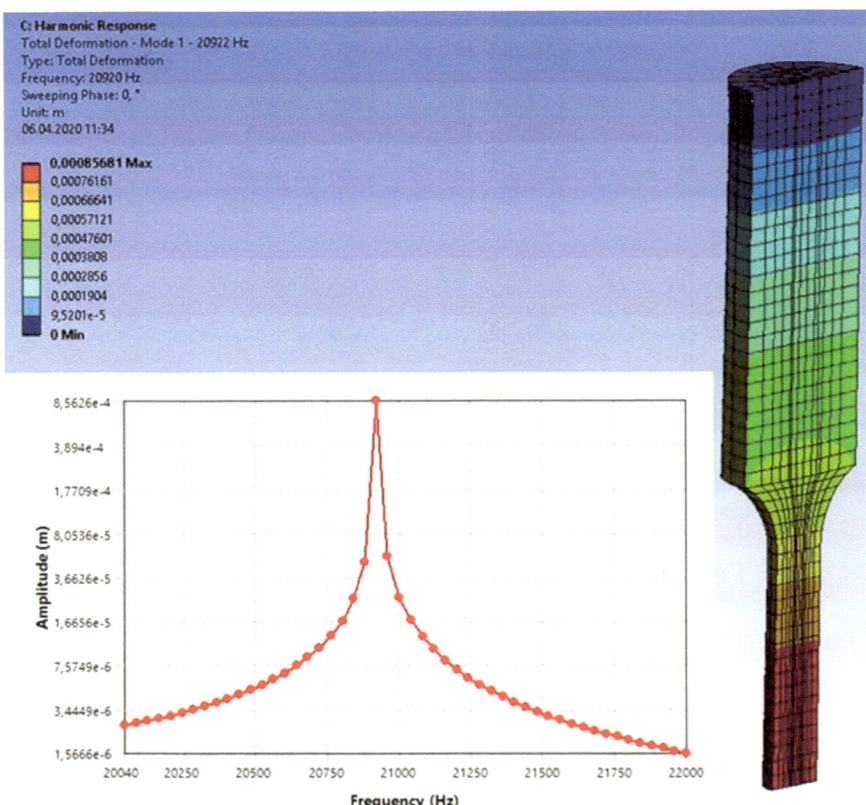

Fig. 3.20 Resonance of the horn itself

analogues. A decrease in the specific capacity during cycling of the GGC at a current density of 100 mAg^{-1} after 95 cycles has not been noted. Exceptionally low decrease in the specific capacity upon the increase of the power load, perfect cycling stability, and high Coulombic efficiency supported by electrochemical impedance analysis indicate good prospects of using GGC as a lithium-ion battery anode and for utilizing graphene additives to electrode materials of lithium-ion batteries operating at high discharge currents.

In past decades, lithium-ion batteries (LIBs) have become important energy storage devices indispensable in modern society, from portable electronics like mobile phones and laptops to electric vehicles, power grids, and military applications. This is in line with the existing trend of gradual switching-over from fossil fuels to sustainable energy resources including solar, wind, and geothermal energies.

Increasing industrial and domestic applications require improving LIBs performance, i.e., increasing their capacity and shortening charge/discharge time.

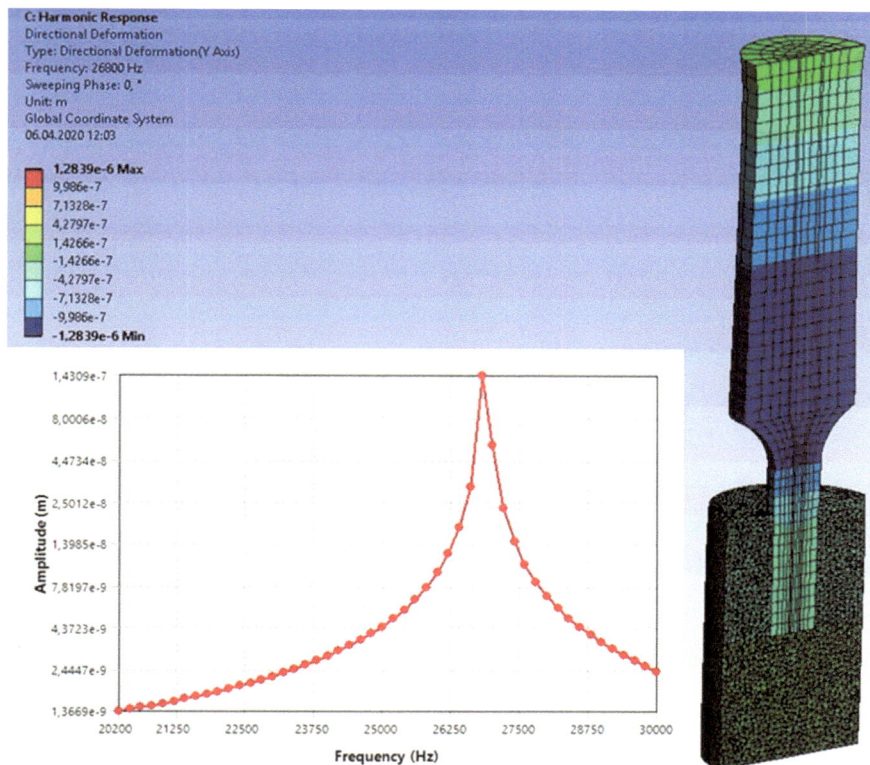

Fig. 3.21 Resonance of a horn + liquid

To reach these goals, composites containing two-dimensional (2D) materials based on graphenes may be of great significance [17–19]. LIBs consist of two intercalation electrodes both able to reversibly insert/deinsert lithium ions accommodated in respective positions of a crystal lattice [20]. An anode is usually made of carbonaceous materials (or sometimes of L i4Ti$_5$O$_{12}$ or TiO$_2$), and a cathode is represented by oxide materials—based on transitional metal ions (Co^{+3}, Ni^{+3}, Mn^{+3}, Mn^{+4}, etc.), which change their charge upon charging/discharging the battery.

Among advantages of LIBs is their high operating potential determining high specific energy (energy density). Intrinsic drawbacks of intercalation electrode materials are also well known as mainly caused by their crystalline, three-dimensional nature [20, 21]. For example, high-rate properties of LIBs are limited by the diffusion of lithium ions in structural channels or interplanar interstices in the bulk of the electrode material and hence meet significant spatial hindrances. Due to this fact, charge/ discharge rate (specific power and power density) of LIBs is often limited. Clearly, such a restriction is much weaker in 2D graphenes where the diffusion of lithium ions occurs primarily on the surface of a graphene plane. This means that

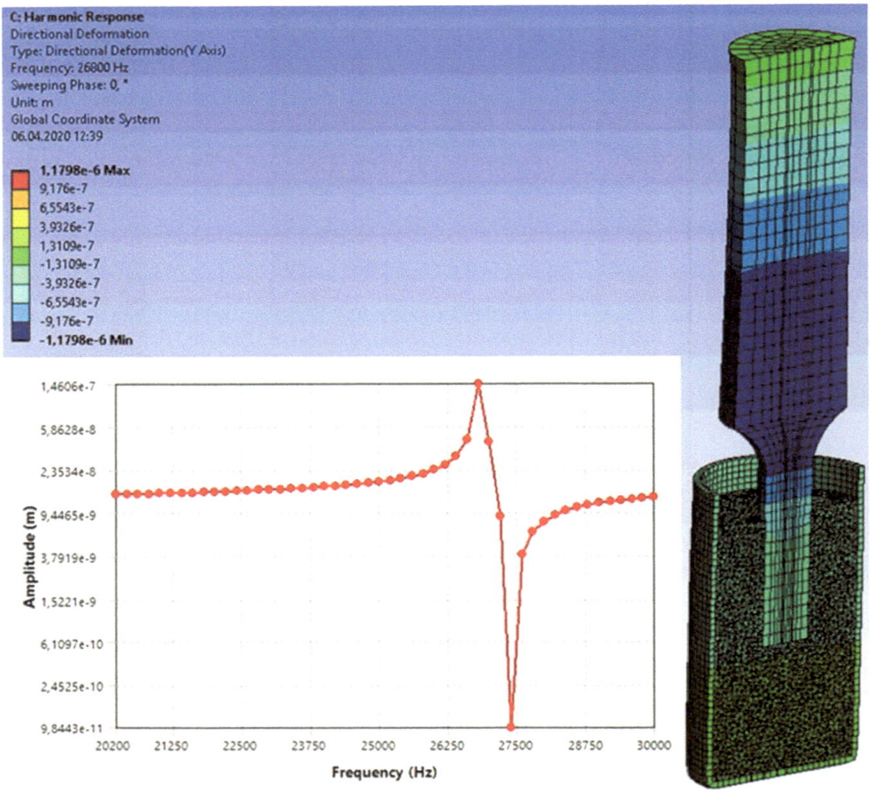

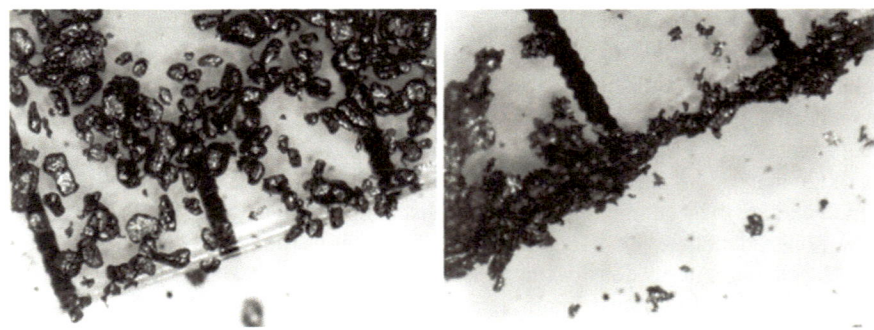

Fig. 3.22 Resonance of a horn + liquid (frequency—26,750 Hz)

Fig. 3.23 Close-up of TEG before and after ultrasonic treatment

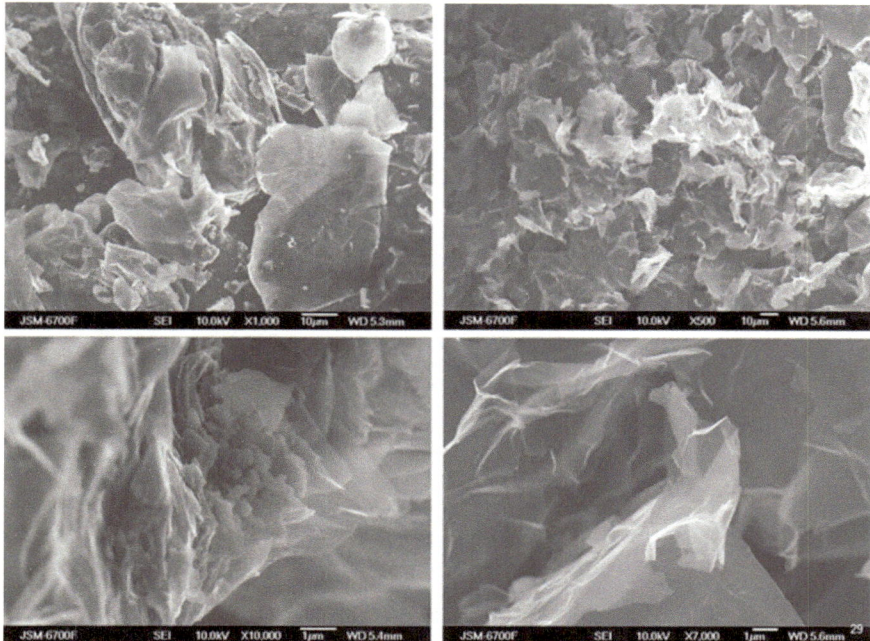

Fig. 3.24 Photomicrographs of TEG before and after ultrasonic treatment

Fig. 3.25 Water hammer with the Yutkin effect

large surface area as well as high conductivity of graphenes and their composites can lead to increasing charge/discharge rate of LIBs. Among other advantages of graphenes are higher specific capacities than those of ordinary graphites. Just these two circumstances reveal that using graphenes and/or their composites as electrodes in LIBs can greatly improve their overall performance and power density on the first place.

Fig. 3.26 Electrohydraulic
shock

In our recent paper [22], obtaining of globular graphene based on thermally
expanded natural graphite has been described. The aim of this work is to study
the electrochemical properties of a graphite-graphene composite (GGC) synthesized
in this way and to determine further prospects of its application as an anode in high
energy density lithium-ion batteries.

Graphene-containing samples were obtained from natural graphite as discussed
in our previous work [22]. Samples of natural graphite were oxidized by concentrated sulfuric acid and further treated by the mixture of $KMnO_4$ and H_2O_2. After
this, oxidized graphite was subjected to a thermal shock at 800–900 °C resulting in
thermally expanded graphite [23]. This material was mixed with water at 1:3 ratio,
first dispersed and compacted in a homemade mechanical cavitator and then finally

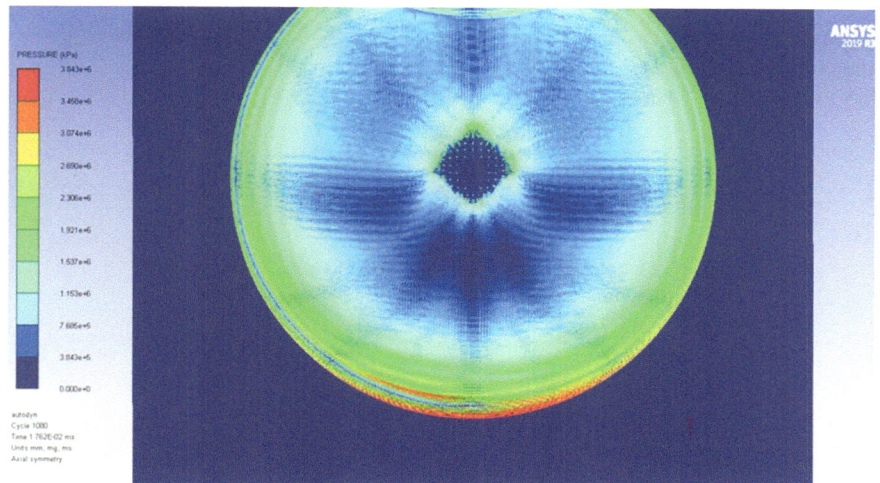

Fig. 3.27 Simulation of an underwater explosion without taking into account the reaction of the tank walls. Pressure in the environment of 3.8 MPa (at the time of 0.0176 ms)

dispersed in an ultrasonic disperser (UZDN-1200, Ukraine) at 1200 W power at 30 kHz frequency. This approach significantly simplifies the synthetic procedure of GGC samples, reduces specific energy costs, and might be prospective for scaling.

Electron micrographs were taken on MIRA TESCAN (Czech Republic) and JSM-6700F (JEOL, Japan) electron microscopes. Raman spectra were excited by 633 nm radiation and recorded with a resolution of 1 cm^{-1} on an inVia Renishaw Raman microscope (UK).

Electrochemical studies were performed on a MTech PGP-550 M/S potentiostat–galvanostat (Ukraine) in 2016 coin cells with a lithium metal anode serving as a counter and reference electrode. To prepare working electrodes, materials in question and a NV-1 T water binder with the dry component ratio of 85:15(without additional carbon black content) were suspended in distilled water and mixed in an RW 20(IKA, Germany) high-speed mixer. Similar binders, as found in our recent work [24], are much advantageous for increasing the performance of LIBs. Resulting slurries were cast onto an aluminum foil with a doctor blade and dried under an IR radiator. The quantity of graphite-graphene composite in a dry remainder was of 4–5 mg cm^{-2}. Electrodes were divided by a Celgard 2500 separator membrane impregnated with a 1 mol L^{-1} solution of LiPF$_6$ in a mixture of ethylene carbonate/ethyl methyl carbonate (1:1 by mass). Cell assembling was made in a dry glove box. Electrochemical impedance spectra (EIS) were measured using an Autolab electrochemical station (Netherlands) in the frequency range of 10^6–10^{-1} Hz at the voltage of 3.0 and 0.01 V. The oscillation amplitude was 5 mV.

Physicochemical Properties. The term "graphene" refers primarily to a carbon particle containing a single graphene layer. Graphene particles with two or more

Fig. 3.28 Simulation of an underwater explosion and its reflection from the walls of the vessel (Plexiglas 5 mm)

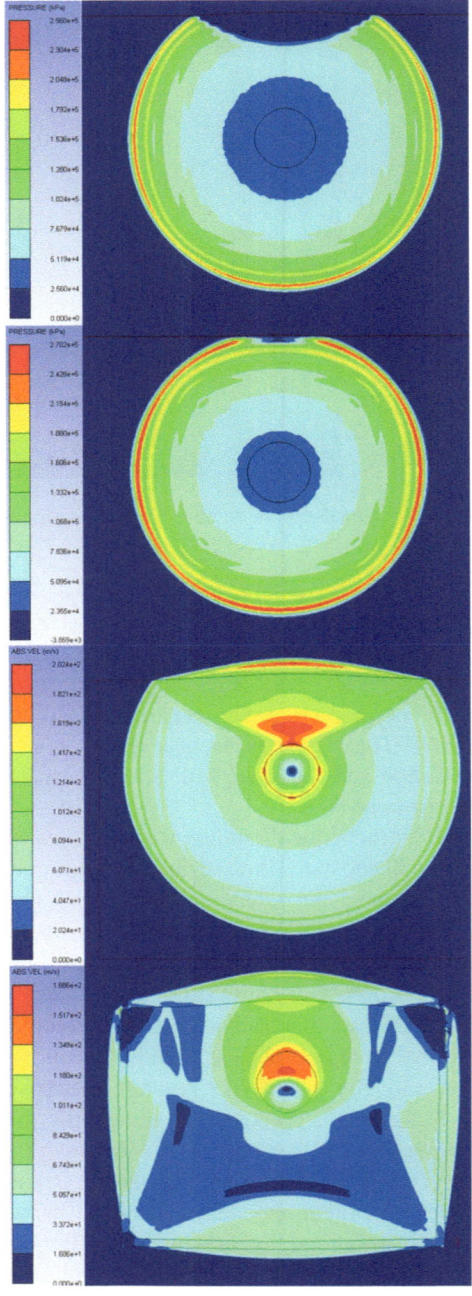

layers are described, and their properties are determined by the number of layers and not by the graphite macroparticle as a whole. The upper limit of the number of layers of graphene is set by a function of properties different from those of graphite. As such limit is established in each specific case of the experiment, many researchers approach the graphene problem from a formal point of view and confine themselves with systems of 100–150 layers. Another point of view associates the term "graphene" with an arrangement of a material and includes inconsideration mixtures of graphenes with a different number of layers, sometimes mixed with graphite. A recommended nomenclature for two-dimensional carbon materials may be found in [25].

The samples were studied by the SEM method. At a small magnification, globules with a size of 50–150 μm were observed (Fig. 3.29a). Magnification up to 106 times made it possible to detect graphene layers that do not belong to crystalline graphite (Fig. 3.29b, c). As a result of the twisting of the particles, it is only sometimes possible to determine the number of layers in the images (Fig. 3.29d) where three-layer graphene was discovered. It is obvious that the material belongs to the nanoscale ones, having a size of up to 100 nm in one of the dimensions.

Fig. 3.29 SEM images of the graphite-graphene composite

Raman spectra of the GGC and graphite are presented in Fig. 3.30. For graphites and graphenes, the so-called D, G, and 2D bands [26, 27] are showing up in Raman spectra at ~ 1350, ~ 1580, and ~ 2300 cm^{-1}, respectively. The D band is associated with structural disorder (structural defects), and the G band arises from stretching vibrations of sp^2 bonds corresponding to the ordered carbon. The presence of a distinct D band and a relatively large ID/IG intensity ratio is typical of graphene-containing structural defects. It has been found in our case that in comparison with graphite, the intensity of the 2D peak is greater and the intensity of the D peak is lower in the graphene array (Fig. 3.30). This is considered characteristic of graphenes when they are formed from graphite. The 2D band of the GGC sample has an asymmetric profile with a maximum at 2736 cm^{-1}. The position of the maximum is shifted to the short-wavelength region compared to highly crystalline graphite at 2747 cm^{-1}, whereas single-layer graphene has a band at 2717 cm^{-1}. It is obvious that the obtained sample mainly contains multilayer graphenes mixed with thermally expanded graphite.

To more accurately determine the number of sheets in graphene arrays in the GGC samples, we have compared our data with the model spectra presented by the discoverers of graphene in one of the most cited papers in the field [26], where the Raman profile of the 2D peak is suggested for this purpose. The profile shown in curve 1, Fig. 3.30, is most close to that displayed in Fig. 3.30c of [26] for five-layered graphene. This reveals that graphene arrays in our material consist of no more than five graphene layers. Such conclusion is quite close to our estimate made from SEM data.

Electrochemical Measurements. While the theoretical capacity of graphite equals to 372 mAhg^{-1} being attributed to the formation of one-sided LiC$_6$ structures in the graphite body, the theoretical capacity of graphenes and graphene-like carbonaceous materials is for long known to far overwhelm it and reaches 744 mAhg^{-1} due to

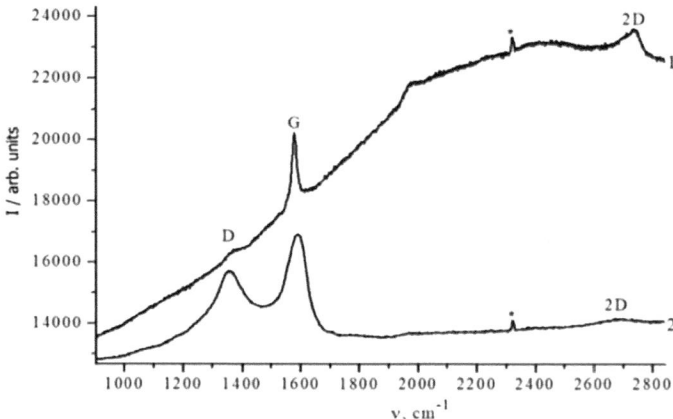

Fig. 3.30 Raman spectra of the graphite-graphene composite (1) and graphite (2)

the formation of two-sided LiC$_3$ aggregates. Furthermore, this is not a limit and, experimentally, the final specific capacity value for graphene-like systems can be extended up to 1200 mAhg^{-1} [28]. Such a phenomenon may be attributed to defects generated in the course of the preparation of graphenes and accommodating extra lithium ions in them [29].

Results of electrochemical tests of the graphite-graphene composite in the voltage range of 0.01–3.0 V (cycles 1–5) are shown in Fig. 3.31. During the intercalation of lithium ions in GGC, a plateau at 0.8 V (relative to Li+/Li) associated with the formation of a solid electrolyte interface (SEI) on the electrode surface is observed in the first cycle. It should be noted that, as seen from Fig. 3.31, the value of the irreversible capacity in the first cycle is ~ 50% and may be related to the specific surface area of the electrode. With further cycling, the Coulombic efficiency determined as the ratio of discharge to charge specific capacity increases to 99.5%. Upon cycling, one can observe the presence of three plateaus on the galvanostatic charge/discharge curves of GGC, which correspond to the stepwise intercalation/deintercalation of lithium ions into/from the material. The quasi-stationary discharge capacity of the graphite-graphene composite amounts at 290 mAh g^{-1}. This is close to the lowermost existing values reported in the literature for graphene-containing samples [30]. An increase in voltage above 0.3 V upon deintercalation of lithium ions leads to a rapid increase in voltage up to 3.0 V. These circumstances may indicate the presence of a not-so-great amount of graphene in GGC.

Not so much attention was paid to high-rate properties of graphene-like materials [28–30], possibly, because of their much higher specific capacities than those of common cathode materials [20, 21]. Taking into account that graphenes have high specific conductivity values, we carried out power tests of the synthesized GGC (Fig. 3.32). The intercalation of lithium ions into the GGC was performed at a constant current density of 50 mA g^{-1}, in order to avoid an ohmic voltage "dip" leading to the formation of metallic lithium on the surface of the material. It should be stressed that 20-fold increase in the current density (from 50 to 1000 mA g^{-1})

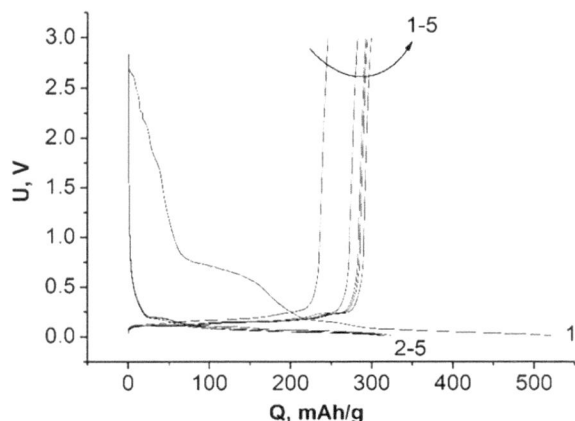

Fig. 3.31 Galvanostatic curves of the GGC. The current density is 50 mA g^{-1}

Fig. 3.32 Changes in the
specific capacity of the GGC
depending on the current
density of deintercalation of
lithium ions

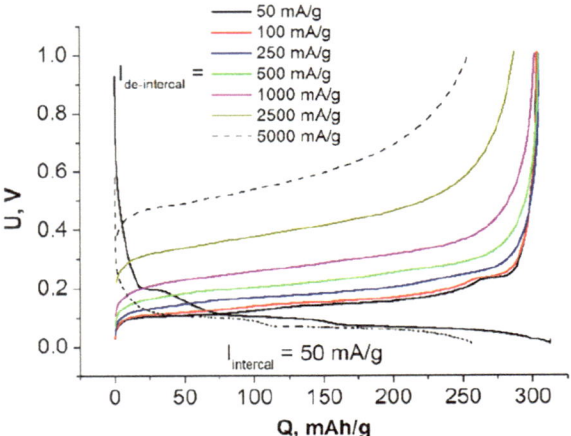

Fig. 3.32 Changes in the specific capacity of the GGC depending on the current density of deintercalation of lithium ions

leads to almost no changes in the specific capacity upon deintercalation of lithium ions. This feature favorably differs the material studied from existing analogues which demonstrate fast capacity decay at corresponding current loads [30] and may be advantageous for future applications. A further increase in the current density leads to a decrease in the values of the specific capacity during both the intercalation and deintercalation processes of lithium ions.

Due to the fact that above 1 V, as has been shown earlier (cycles 1–5), no processes are observed; the cycling of GGC was limited to a voltage range of 0.01–1.0 V matching the operation voltage of a graphite electrode. Lithium ions have been deintercalated from the GGC at varying current densities. The dependence of the specific capacity of GGC on the cycle number is shown in Fig. 3.33. The first 40 cycles are power tests of the sample discussed in Fig. 3.32. Cycles from 41 to 95 are galvanostatic resource cycling results at a constant current density of 100 mA g^{-1} demonstrating the perfect cycling stability of the sample. The specific capacity of GGC upon cycling is kept constant at 290 mAh g^{-1}, and the Coulombic efficiency with a lithium counter electrode is 97.5%. Together with exceptionally low decrease in the specific capacity upon the increase of the power load, this indicates the good prospects of using GGC as a LIB anode.

In order to gain insight in excellent high-rate properties of the material studied, EIS spectra have been measured. Nyquist diagrams and equivalent schemes for calculating the parameters of the electrochemical impedance of the GGC are shown in Fig. 3.34. The resistance values Re corresponding to the resistance of the electrolyte in the pores of the separator in both cases slightly differ from each other and amount to 4.8 ± 0.2 Ω. However, depending on the state of the electrode, different shapes of the impedance spectra are observed, which, in turn, may be described by different equivalent circuits. Thus, at a voltage of 3.0 V, the Nyquist diagram demonstrates one semicircle with the center below the abscissa axis, which corresponds to the process of charge transfer through the SEI film on the lithium electrode and an inclined

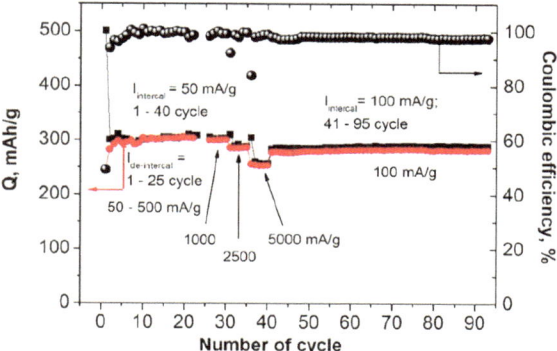

Fig. 3.33 Changes in specific capacity and Coulombic efficiency upon cycling of the GGC. Voltage range is 0.01–1.0 V

straight line (Warburg impedance), which describes the processes of intercalation of lithium ions into the GGC. In this case, instead of the Warburg impedance in the equivalent circuit, a constant phase element CPE is used, which more accurately describes the ongoing processes (since both intercalation processes and lithium-ion adsorption processes are observed in GGC, which is expressed in adding a capacitive component to "classical" Warburg impedance) resulting in the deviation of the slope of the straight line from 45°. The value of charge transfer resistance through the SEI film Rct is $19.1 \pm 0.2\ \Omega$.

In contrast to the previous picture, two semicircles are showing up in the diagram at 0.01 V, corresponding to the processes of charge transfer through the SEI on the GGC and lithium electrodes. The resistance value Rct is the sum of the resistance of two semicircles and equals to $29.3 \pm 0.2\ \Omega$. In addition to the presence of the 2nd semicircle, there are differences in the slope of the linear part at low frequencies. The increase in the SEI resistance during the intercalation of lithium ions in the region of low voltages is explained by the fact that most of the vacancies for the intercalation of ions are already occupied and mutual repulsion can occur between them.

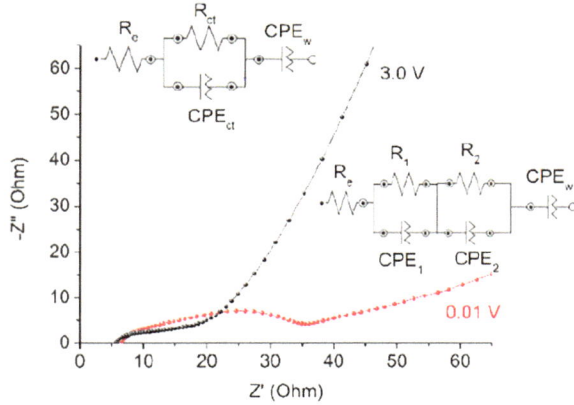

Fig. 3.34 Nyquist diagrams and equivalent schemes for calculating the resistance values

Summary

Graphite-graphene composites have been obtained as a result of high-speed heating of oxidized graphite. Raman spectroscopy proves the presence of ordered graphene in graphite-graphene composites as evidenced by the ratio of the intensities of the D and G peaks. The predominant formation of no more than 5 graphene sheets in the material as concluded by the analysis of the profile of 2D peak is supported by SEM micrographs.

Electrochemical tests of GGC samples show that in spite of quite low specific discharge capacity (290 mAhg^{-1}), a 20-fold increase in current density (from 50 to 1000 mAg^{-1}) does not lead to a change in the specific capacity upon deintercalation of lithium ions. This feature favorably differs the material studied from existing analogues. A decrease in the specific capacity during cycling of the GGC at a current density of 100 mAg^{-1} after 95 cycles has not been noted. Exceptionally low decrease in the specific capacity upon the increase of the power load, perfect cycling stability, and high Coulombic efficiency supported by electrochemical impedance analysis indicate good prospects of using GGC as a LIB anode and for utilizing graphene additives to electrode materials of LIBs operating at high discharge currents.

3.4 Nanolayered Graphite Seals of Extremely High Durability for Nuclear Reactors

The problem of leak proofness of the plug-type connections of reactor plants has been among the most urgent ones in designing, manufacturing, operating, and repairing NPP equipment. Depressurization can lead to accidents involving the release of radioactive coolant outside the corresponding loop of the reactor plant, as well as cause corrosion damage to the reactor plant structural elements. The strength and tightness of sealers is a precondition for trouble free operation of reactor plants.

Traditionally, almost all plug-type connections were sealed with nickel rings. As ring is compressed, nickel, due to its properties, hardens as result of strain and becomes as hard as the sealing surfaces of plug-type joint flanges, which are made mainly of austenitic steel 08X18 H10T or have a surfacing made of this steel. Tightening results in plastic deformation of the sealing surfaces, which changes their geometry. Over the years, these processes are intensified, plastic deformations are accumulated, and the seal assembly loses its tightness. The sealing surfaces need to be repaired to restore their design geometry, but this does not always lead to a positive result. In the early 1990s, leakages of plug-type connections of WWPR (water-water power reactor) plants occurred massively, which led to NPP forced outages until the joints were resealed. The technological principle of manufacturing the sealing elements made of graphite foil obtained by thermally expanded graphite (TEG) rolling is based on the results of research carried out at joint Stock company research design and aiming at upgrading seal assemblies of steam generators PGV-440 and PGV1000 M [31–39].

Five samples of TEG foil from three manufacturers have been used while studying the technology for compression of TEG sealing elements, at the Institute of Gas of the N.A.S of Ukraine:

- FG-IG graphite foil manufactured at the Institute of Gas of the N.A.S of Ukraine. The foil is made of TEG made at and using the technological equipment of the Institute of Gas of the N.A.S of Ukraine. TEG is obtained from oxidized graphite of Zavalievsky Graphite Plant, using the technological equipment of the same institute. the technology for obtaining TEG is described in [40, 41];
- TMG-F/V2 graphite foil manufactured by LLC *TmSpetsmash* (Kyiv, Ukraine) made of TEG of its own production, by rolling, without binder; TEG is obtained from oxidized graphite produced in china;
- GF-1, GF-2, GF-3 graphite foils supplied by BoNUM GroUP LLC (Zaporizhzhia, Ukraine) and manufactured by *Yichang Xincheng Graphite Co.*, Ltd., China, meet the requirements of the XC-120109 standard.

The physical, technical, and physicochemical characteristics of graphite foil GF measured in the course of study are given in Table 3.1.

Experimental samples of gaskets are made using press molds of standard sizes (Figs. 3.35 and 3.36).

The graphite foil is wound into rolls on inserts having a diameter over 90 mm. The roll length must exceed the width of the foil, at least, by 10% the roll mass is not regulated. Before pressing, the graphite foil is cut into strips having a width of 20–25 mm, with the total length of the strip segments corresponding to a given mass of gasket based on the density of the gasket material (1.80 g/cm^3).

The graphite foil strips of appropriate length are wound around the rod of press mold manually, keeping the required winding density further, a certain mechanical method for winding will be provided for molds having dimensions meeting the specification requirements.

For pressing the sealing gaskets, hydraulic press with a working table of required dimensions and a compression force indicator is used and the specific compression force is 150 MPa. After the end of compression, the gasket is held in the press mold, in compressed state for 5–10 additional seconds to prevent a high rebound after the release of pressure. The finished gasket is treated with fine-grained abrasive material to remove scratches and tears.

For each type of source material, 15 gasket samples are made, five of which are used for measuring the ultimate compressive strength, and ten samples are used to determine the density characteristics of gaskets.

The ultimate compression strength of TEG gasket prototypes is measured in accordance with TU 5728-006-12058737-2005 "sealing gaskets of thermally expanded graphite". The method provides for applying the axial load to the prototypes located between two parallel supports, until the prototype is destroyed. The tests are carried out at a temperature of 20 °C. The gasket is placed between the two supports. The load is applied smoothly (without shocks) until the prototype is destroyed. The maximum load (Q) recorded before the prototype is destroyed is used to determine the ultimate compression strength.

Table 3.1 Physical, technical, and chemical properties of graphite foil samples used in experiments for improving the technology for manufacturing sealing gaskets

Parameter	Type of graphite foil				
	FG-1G	TMG-F/V2	GF-1	GF-2	GF-3
Appearance	TU U26.8-30969031-002-2002[a]	TU U26.8-30969031-002-2002[a]	TU U26.8-30969031-002-2002[a]	TU U26.8-30969031-002-2002[a]	TU U26.8-30969031-002-2002[a]
Thickness, mm	0.19–0.22	0.42	0.51	0.53	0.21
Density, g/cm^3	0.98	1.01	1.0	0.82	0.91
Carbon content, wt.%	99.0	99.3	98.7	97.8	97.7
Ash residue content, wt.%	0.07	0.44	0.84	1.9	1.9
Sulfur content, wt.%	0.10	0.11	0.12	0.12	0.12
Tensile strength along the rolling axis, MPa[b]	3.3–4.2	4.2–5.0	3.7–3.0	3.3–3.5	3.5–3.9
Tensile strength across the rolling axis, MPa[b]	2.7–2.9	4.0–4.9	3.4–3.7	3.0–3.2	3.1–3.9

(continued)

Table 3.1 (continued)

Parameter	Type of graphite foil				
	FG-1G	TMG-F/V2	GF-1	GF-2	GF-3
Electric resistance along the rolling axis, ohm m	7.1×10^{-6}	8.2×10^{-6}	6.6×10^{-6}	7.4×10^{-6}	7.1×10^{-6}
Compressive ratio in a free state, under a pressure of 35 MPa, %	37.0	36.0	34.0	37.0	35.0
Restorability after release of pressure of 35 MPa, %	11.0	14.0	12.0	15.0	14.0

[a] foil surface is flat, smooth, having neither nicks, nor scratches, nor ruptures, nor burns, nor buckles, nor laps, nor corrugations, nor foreign impurities; [b] limiting values obtained for five samples

Fig. 3.35 Mold prototype of standard size 1: 1—outer chase; 2—supporting ring; 3—inner chase; 4—plug; 5—gasket

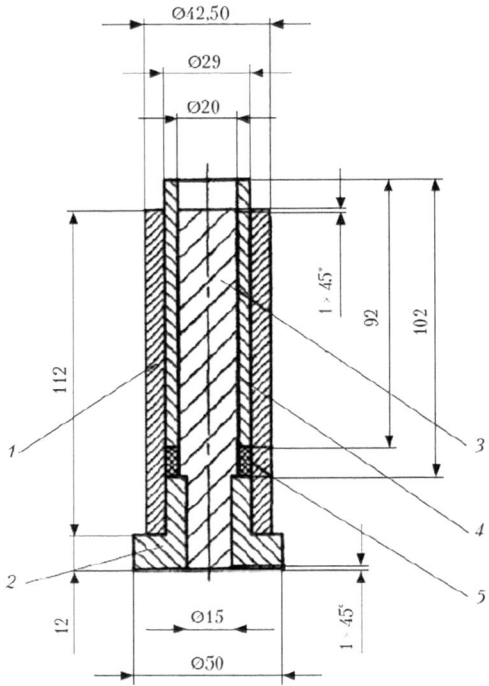

The ultimate compression strength σ_{st} in MPa is calculated by the formula:

$$\sigma_{st} = \frac{Q}{F_0},$$

where Q is maximum load recorded before the prototype destruction, N; F_0 is prototype's working surface area, mm^2.

The ultimate compression strength value has been determined as arithmetic mean of tests of five prototypes. The test results are given in Table 3.2.

The ultimate compression strength of *HYDROPRESS* gasket prototype is 9.5 MPa, which ensures the proper density of plug-type connections of NPP technological equipment.

The aforementioned research has been carried out at the research base of the Institute of Gas of the N.A.S of Ukraine, using the existing technological equipment. The tests at a pressure of 350 kg/cm^2 have been done on the test stand of *ATOM-PRILAD* design Bureau of *Energoatom* National Atomic Energy Company. The mold for making TEG gaskets for steam generator PGV-1000 M has been designed and manufactured at the Brovary Plant of Powder Metallurgy. The gasket prototype manufactured at the same plant completely fits the seal assembly of the steam generator. This means the design is based on the correct assumption of the gasket expansion

Fig. 3.36 Mold prototype of standard size 2

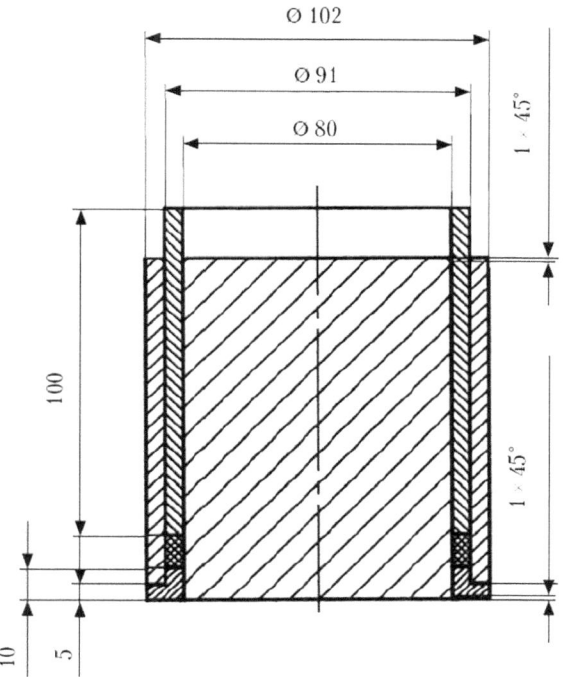

Table 3.2 Ultimate compression strength of TEG gasket prototypes

№	Parameter	Type of graphite foil				
		FG-1G	TMG-F/V2	GF-1	GF-2	GF-3
1	Ultimate compression (σ_{st}), of standard size 1 gasket, MPa	6.47	6.20	5.46	6.76	6.45
2	Ultimate compression strength (σ_{st}), of standard size 1 gasket, MPa	6.09	6.11	5.77	6.55	6.77

after its removal from the mold. Currently, at *Atomenergomash*, where the necessary production facilities and equipment are available, there are created a complete cycle for manufacturing the TEG gaskets. In the test mode of production site operation, several samples of sealing gaskets having standard dimensions have been made.

According to the research results, a technology for manufacturing the TEG sealing gaskets by pressing a semifinished product and the graphite foil has been developed. The physical and chemical characteristics of graphite foil and their influence on the properties of finished products have been studied.

The tests of TEG gasket prototypes have shown that their mechanical strength is 30–35% lower than the strength of the previously used TEG gaskets. It has been established [42] that pre-treatment of the derivative graphite foil by increasing the roughness of its surface enables a 20–30% increase in the mechanical strength of the gaskets.

While manufacturing the gasket prototype of standard size for the steam generator PGV-213, it has been established that the gasket expansion after its removal from the mold corresponds to the results of previous studies.

Today, the Institute of Gas of the N.A.S of Ukraine continues research works to improve the mechanical strength of gaskets by pre-treatment of graphite foil, in particular, by increasing the surface roughness. Based on the results, the patent of Ukraine for invention [42] has been received and two applications for the patent of Ukraine have been filed.

The research results are used to create a production site for manufacturing TEG gaskets at one of *Energoatom* enterprises, *Atomenergomash*. The implementation of the entire production cycle for manufacturing the TEG sealing gaskets will eliminate the import dependence of the country and increase the safety of the operation of domestic nuclear power plants.

Summary

Within the framework of the conducted R&D works, the physical, technical, and physicochemical properties of graphite foils made of natural graphite derivatives by various manufacturers have been established; molds of two standard sizes have been designed and manufactured; the technology for molding the sealers from graphite foil has been tested; specific compression force to obtain the sealers of given density has been determined; specific compression strength of the obtained gaskets has been established.

3.5 Technology of Water Cleaning from Oil Products with the Help of Graphite Supersorbent

Emergency spills of oil and oil products happen quite often and in different scales. Elaboration of effective ways to eliminate spilled oil and oil products are very actually now days. One of the most effective methods for solving this problem is the absorption of the oil by sorbents. Nanoflaky oil supersorbent on the basis of expanded graphite has unique characteristics. It is a special modification of the graphite obtained by multistage thermo-chemical treatment of natural graphite. There were developed different methods of preliminary preparation and subsequent application of this sorbent considering a specificity of emergency spill, properties of adsorbed liquid, the nature of cleaning surface, and weather conditions. Also, question of spent sorbent utilization has been studied. The recycling process involves the desorption process up to 85% with subsequent use of the sorbent to ten cycles of regeneration. Obtained while desorption liquid can be used for another purpose or as an additive to fuel oil. Technology and equipment for liquidation of emergency oil spills on water surface and coastal sand have been developed and tested.

At the present time for elimination pollutions by oil, petroleum and other organic fluids on the water surfaces, sand beaches, as rule porous substances of natural and artificial origin such as: peat, sawdust, shredded twigs, perlite, polystyrene foam, various fibrous materials are used. Sorbents are applied to the contaminated area of the water and sand surface after that the major part of spilled product has been collected more often by mechanical means. Also, it was proposed to use special bacteria that decompose organic matter into the neutral substance [43].

At liquidation of emergency spills of oil and oil products by the method of sorption the most promising method is the use as oil-absorbing sorbent thermoexpanded graphite (TEG). TEG represents itself a special modification of the graphite obtained by multistage natural graphite chemicothermal processing. This kind of graphite is found in the literature also under the names of exfoliated graphite, foamed graphite, and thermografenite. This product is characterized by very low bulk density (2–5 kg/m^3) and high specific surface which in combination with its oil-receptivity causes a high absorption capacity relative to oil and other hydrophobic organic liquids. One gram of this substance can absorb 30–60 g of oil (see Table 3.3). An important feature of this sorbent is its inertness, the ability to desorption up to 90% of absorbed liquids, and the possibility of thermo-chemical regeneration for repeated use [44–46].

Actually liquidation process of emergency spills of oil, oil products, and other organic liquids on the water surfaces and sand of coastal zone with sorbent on the basis of thermoexpanded graphite includes such stages:

- sorbent obtaining—thermoexpanded graphite (if it required directly on the place of emergency spill);
- pre-treatment (preparation) of the sorbent;
- applying of a sorbent on contaminated surface;
- collecting of a saturated sorbent;
- separation and recycling of on absorbed liquid;
- regeneration of a waste sorbent and its reuse.

Table 3.3 TEG sorption characteristics for some hydrophobic organic liquids

The name of the substance	Sorption capacity, g/g sorbent
Acetone	30
Turpentine	30
Benzene	35
Diesel fuel	40
Kerosene	40
Vegetable oil	45
Machine oil	50
Crude oil	55

Scientists from Gas Institute of National Academy of Sciences of Ukraine have developed the technology for thermoexpanded graphite producing. Also they have designed, manufactured, and tested series of installations of different performance, autonomy, and automation to produce TEG.

- Pilot installation with a capacity of 8.5 m³/h (35 kg/h by raw material). One example of the unit was manufactured and put into operation at the Argonne National Laboratory, Chicago, USA (Fig. 3.37);
- autonomous automatic aggregate of a local destination with capacity of 1–2 m³/h (5 kg/h by raw material (Fig. 3.38);
- TEG autonomous knapsack generator with a capacity of 0.8 m³/h (3 kg/h by raw material) (Fig. 3.39) [47];
- Fig. 3.40 demonstrates the simulation of a mobile unit for TEG production in s container mounted on a wheeled chassis (truck GAZ-66-02).

The last three types of installations can work offline and be used in liquidation of oil spills as local and in remote places. Significant technical obstacle when using oil-absorbing sorbent on the basis of expanded graphite is its extremely low bulk density, resulting in low profitability of the technology as a whole with TEG delivery to the place of emergency spill. Spreading of the sorbent in the form of a dry powder to the contaminated surface is also associated with its entrainment (losses), which causes an increase in specific consumption of sorbent and contamination of the surrounding area.

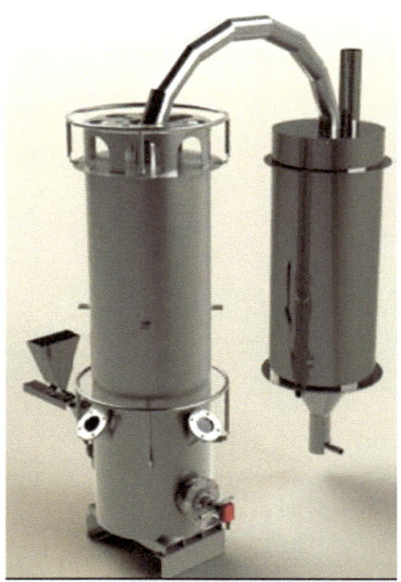

a   *b*

Fig. 3.37 Pilot unit for TEG generation with a capacity of 8.5 m³/h: **a** simulation, **b** in metal

Fig. 3.38 Autonomous
automatic aggregate of a
local destination with
capacity of 1–2 m^3/h by TEG

Fig. 3.39 TEG autonomous
knapsack generator with a
capacity of 0.8 m^3/h (3 kg/h
by raw material)

In the Gas Institute of NAS of Ukraine, various methods of preliminary treatment of the initial TEG have been investigated and tested on a pilot scale. Different modifications of the sorbent have been elaborated.

They are.

- granulation using a binder and the subsequent application of granules by mechanical method (Fig. 3.41) [48, 49];
- pressing to obtain the sorption elements using a binder and without it as well as using reinforcing interlaying and without it, followed by applying mechanical methods (Fig. 3.42) [50–52];

Fig. 3.40 Simulation of a mobile unit in s container mounted on a wheeled chassis (truck GAZ-66-02)

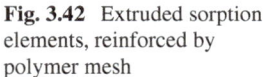

Fig. 3.41 Granular adsorbent of expanded graphite obtained by extrusion (**a**) and clumping (**b**)

Fig. 3.42 Extruded sorption
elements, reinforced by
polymer mesh

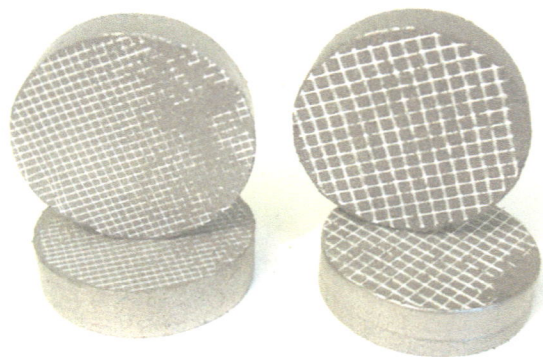

Fig. 3.43 Experimental unit for the preparation and application of a graphite-aqueous suspension

- preparation of the water-graphite suspension with subsequent application by means of a centrifugal pump (Fig. 3.43) [53, 54];
- preparation of the water-graphite foam suspense followed by the application of air-foam jetting [55].

Choosing the method of sorbent preparation on the basis of TEG is accomplished taking into account a specificity of an emergency spill, properties of adsorbed liquid, the nature of the polluted surface, and surrounding area conditions (e.g., availability of water sources for the preparation of water-graphite suspension). It should be noted that the sorption capacity of any sorbent modification obtained after pre-processing of the original TEG is lower than sorption capacity of dry powder due to partial mechanical destruction of nanolevel structure of expanded graphite. For example, at granulation of TEG powder with the use as a binder 2.5% solution of the glue— "PVA", sorption capacity of the pellets, sufficient for reliable manipulation strength (0.15 kg/cm^2) is 23.9 g/g for diesel fuel, i.e., 40% lower than the original TEG. Sorption capacity of pressed sorption elements with density of 12.5 kg/cm^3 is 12– 14 g/g, i.e., three times less than that of the original TEG. However, considering the high degree of adaptability of cleaning operation of contaminated area on the whole a significant reduction of costs for sorbent delivery to the place of an accidental spillage—the economic feasibility of pre-treatment and preparation of initial TEG is justified. A little decrease in sorption capacity compared to original sorption capacity of TEG—8.5–20% is observed when oil-absorbing sorbent from the TEG in the form of water-graphite and foam-graphite suspension is used. Thus, a high degree of cleaning of water surface and coastal sands is provided. Also, pollution of an environment by sorbent is prevented. In addition, this technology is characterized by relative simplicity and does not require the development of special technological equipment. Regardless of the type of pre-processing the original TEG and the method of its application on contaminated surface collecting, the saturated sorbent is produced by any known and proven in practice methods: a perforated material or grid with a mesh size of up to 12 mm [56, 57] or vacuuming [43].

When emergency spillage is small collecting of saturated sorbent can be effected by any suitable means at hand. As noted above, a significant advantage of oil-absorbing sorbent compared to known is the possibility of desorption of absorbed liquid and regeneration of a "pressed" sorbent for reuse [46]. Up to 85% of absorbed oil is separated when desorption occurs by centrifugation. After appropriate treatment, this product can be used for its intended purpose. Developed at the Gas Institute of N.A.S of Ukraine the technology of thermo-chemical regeneration of the spent sorbent provides for a high-temperature treatment of the waste and subjected to a desorption of a sorbent in the furnace of the cyclone type (Fig. 3.44) [58, 59].

Wherein the content of residual oil in the pressed sorbent allows accomplish a thermo-chemical regeneration in the autothermal regime [59]. Experimentally, it was proven a principle possibility of ten regeneration cycles with maintaining an acceptable sorption capacity of the regenerated sorbent (Fig. 3.45).

Summary

Proven technologies and technical solutions can be used as a basis for the creation of technical units in ministries and departments, intended for the effective and rapid elimination of consequences of emergency spills of oil and oil products on water surface and sand.

Fig. 3.44 External view of the experimental unit for the regeneration of a spent sorbent in a furnace of "cyclone" type

Fig. 3.45 Dynamics of
changes in the sorption
capacity of the regenerated
sorbent with increasing
cycles of regeneration

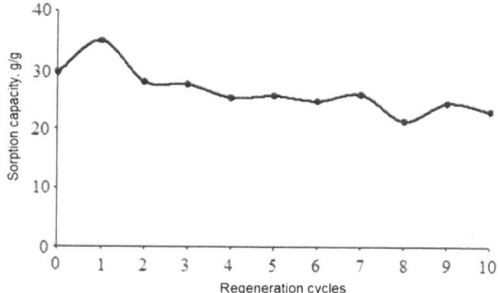

3.6   Structural Features of Carbon Nanomaterials and Determination of Their Thermophysical Properties Using Pulsed Radiation

Abstract. The thermophysical properties of carbon nanomaterials (CNMs), such as thermally expanded graphite, carbon nanotubes, and globular multilayer graphene, are investigated. A series of samples was produced by the technique of one-sided static pressing. By means of the impulse method, the parameters of the thermal conductivity of the samples were determined depending on the conditions of their production. It is found out that the thermal conductivity of the CNMs increases with increasing density and pressure. It is assumed that the mechanisms of heat transfer in these systems can be related to the electron-lattice interaction. Heat transfer can occur by various transport mechanisms, such as collisions or diffusion. Nevertheless, the increase in thermal conductivity can be related to the electronic mechanism.

Modern science connects our future with nanomaterials and nanotechnologies. The widespread implementation of new technologies, including nanotechnologies, into industrial production is achieved by a high level of standardization, which is not possible without the preliminary development of research methods and devices for evaluating new materials. The methods for investigation of carbon nanomaterials and new techniques and tools that allow accurate determination of their structural and thermophysical parameters are developed in many scientific centers and laboratories around the world [1]. In particular, experimental work on the characterization of nanostructured materials is important, along with their manufacture and application. The study of the properties of carbon nanotubes (CNTs), thermally expanded graphite (TEG) and globular multilayer graphene (GMLG), and the development of new materials based on them is a topical task which also includes the development of new hardware for measurements, data processing and representation.

It should be noted that the real experimental characteristics of carbon nanomaterials are much worse than those predicted by theoretical calculations. First of all, this is explained by the structural imperfections in carbon nanomaterials;

however, the problem to a large extent is related to the nature of the tests which are used in research works [60–62]. It is very difficult to prepare test samples without damaging the nanotubes. Another problem is the practical difficulties in representing the experimental results with sufficient resolution.

A sufficiently large amount of material is required when developing materials with given heat insulation properties, which cannot be provided by the modern technologies of graphene production. At the same time, the requirements for graphene-like modifiers are significantly less strict than those for graphene samples that are designed for use in electronics. In particular, one can expect that graphene fragments with large area are not needed for thermal insulation; the use of carbon monolayers is also not necessary. Therefore, increased attention is paid to the development and investigation of high-tech carbon material of a new generation based on thermally expanded graphite. Thermally expanded graphite (it is often called graphite foam or vermicular graphite) is a low-density carbon material that has unique physical and chemical properties: high specific surface area, sufficiently high thermal and chemical resistance, low thermal conductivity, and high porosity. The present work represents the results of investigations of the technology developed for producing carbon nanomaterials; the structural features of these materials are studied, and their thermophysical properties are analyzed.

Currently, there are two approaches to determining the thermal conductivity of materials: stationary and non-stationary. Each of these approaches has its own sources of uncertainty that can affect the accuracy of the measurements. The most reliable method is stationary one, but it can take a lot of time to measure thermal conductivity of one sample. Therefore, the non-stationary method, or the thermal pulse method, is increasingly used for faster measurements. The following factors are essential during the process of impulse heating of material. Since light penetrates into the sample, the sources of optical heating should be volumetric. The energy is released non-uniformly throughout the volume of interaction, as the intensity of light decreases as it penetrates into the sample. This leads to spatially inhomogeneous heating of the material and causes heat and mass transfer processes between different zones of the sample [63].

The aim of the present work was to study the structural features and thermophysical properties of thermally expanded graphite, carbon nanotubes, and globular multilayer graphene.

3.6.1 Thermal Effects Due to Interaction of Pulsed Radiation with Material

The heat transfer in a solid body is determined by the equation of conservation of energy. We can write the amount of heat received per unit volume of material dQ as $dQ = \rho T dS$, where ρ is the density of material; T is the thermodynamic temperature; S is the entropy of unit mass of material. When $dQ = 0$, the heat transfer process is adiabatic. Processes in systems surrounded by a heat-insulating layer, or processes that occur so quickly that heat exchange between the system and its environment does not have time to occur, are considered approximately adiabatic. The adiabaticity of the process is violated in the presence of external sources of heat, in particular optical, and when irreversible processes, such as heat conduction, occur in the system. Energy carriers can be, for example, conduction electrons in metals, phonons in crystalline solids, and light quanta in the case of radiant heat conduction [64, 65].

Let us assume that the thermodynamic parameters change at constant pressure, and thermophysical characteristics, such as the coefficient of thermal conductivity κ, density ρ, and heat capacity at constant pressure c_p do not depend on temperature and coordinates (linear solution of the problem). We can write Fourier's law, according to which the specific flow vector $\vec{q}$ is proportional to the temperature gradient: $\vec{q} = -\kappa \nabla T$. We will assume that each of the pulse beams propagates along the Oz axis and falls normally on the surface (x, y) of the solid body (the z axis is directed normally to the surface vertically downwards) that causes a volumetric heat source in the body with a power density $\alpha J(\vec{r}, t)$, where α is the absorption coefficient for pulsed radiation, and $J(\vec{r}, t)$ is the distribution of light intensity in the material. For the case of a beam with a Gaussian intensity distribution, we can write

$$J(\vec{r}, t) = (1 - R)J_0 \exp(-\alpha z) \exp\left\{-\frac{x^2 + y^2}{r^2}\right\} \cdot f\left(\frac{t}{\tau_p}\right), \qquad (3.1)$$

where J_0 is the intensity of pulsed radiation, R is the coefficient of light reflection by the surface, and r is the radius of the light beam; the function $f(t/\tau_p)$ describes the time contour of a pulse with duration τ_p.

To get basic qualitative ideas about the process of optical heating of materials, it is sufficient in many cases to simulate continuous pulsed radiation with the Heaviside step function:

$$f\left(\frac{t}{\tau_p}\right) = H(t) = \begin{cases} 0, \ t < 0 \\ 1, \ t \geq 0 \end{cases}, \qquad (3.2)$$

and pulsed radiation—by a rectangular contour of intensity:

$$f\left(\frac{t}{t_p}\right) = \frac{H(t) - H(t - t_p)}{t_p}. \qquad (3.3)$$

In the limiting case, when $t_p \to \infty$, we get an instantaneous impulse, which is usually specified using the δ—function, i.e., $f(t/t_p) = \delta(t)$. From a practical point of view, the pulse with intensity rapidly increasing until the moment t_p, and then slowly decreasing is interesting. Such pulses are generated in order to avoid rapid cooling of the sample, and they can be described, for example, by the formula

$$f\left(\frac{t}{t_p}\right) = \begin{cases} \left(\frac{t}{t_p}\right)^n, & t < t_p \\ \exp[-b(t - t_p)], & t > t_p. \end{cases}$$

When the thermophysical parameters do not depend on temperature and coordinates, the temperature distribution during pulsed irradiation is described by a non-stationary linear equation of thermal conductivity:

$$\frac{\partial T}{\partial t} = \chi\left(\frac{\partial^2 T}{\partial x^2} + \frac{\partial^2 T}{\partial y^2} + \frac{\partial^2 T}{\partial z^2}\right) + \alpha * J(x, y, x, t), \tag{3.4}$$

where $\chi = \kappa/\rho c_p$ is the coefficient of thermal conductivity, and $\alpha^* = \alpha/\rho c_p$.

3.6.2 Thermophysical Properties of Carbon Nanomaterials

A carbon-containing gaseous source, natural gas, was used as a raw material for the synthesis of multi-walled carbon nanotubes (MWCNTs). The gas was previously converted on a nickel catalyst, then the conversion products were used for the synthesis of MWCNTs. It was found out that the MWCNTs formed at moderate temperatures due to the treatment of freshly reduced iron with products of air conversion of natural gas [66]. As a result of further crystallization of carbon, nanotubes formed, and iron particles were separated from the main material, i.e., the catalyst material was fragmented. It was found out that carbonization-decarbonization cycles in the $\gamma-$ Fe, $\alpha-$Fe, Fe_3C system play a decisive role in the formation of MWCNTs below 700 °C [66, 67].

In order to determine the dependence of the density of carbon nanomaterials on the pressing force and thermal conductivity, a series of samples was produced by one-sided static pressing in a cylindrical steel matrix with a diameter of 12 mm. The pressing was carried out at a hydraulic press at a pressure in the range of 170–450 MPa. The mass, volume, and density of the samples after pressing were determined. Four samples were produced for each carbon material.

The morphology and microstructure of the synthesized carbon nanomaterials (CNMs) were examined using scanning electron microscopy (SEM) technique at a JSM-6490LV microscope. The microstructure of carbon nanomaterials produced by various methods [68] and synthesis modes is shown in Fig. 3.46.

The CNT material has a so-called sponge morphology (Fig. 3.46a, b), i.e., it is comprised of intertwined tubes of different diameters ranging from 20 to 300 nm. The various diameters are explained by the sizes of the iron grains which detached from the freshly reduced metal plate in a hydrogen atmosphere during the disproportionation of carbon monoxide.

Raman scattering spectra of the samples (see Fig. 3.47a-c) were studied using a mini-RamanPro Raman spectrometer Lightnovo (Denmark). A laser with a wavelength of $\lambda = 785$ nm for the spectral range of 600–2000 cm^{-1} was used. The accuracy

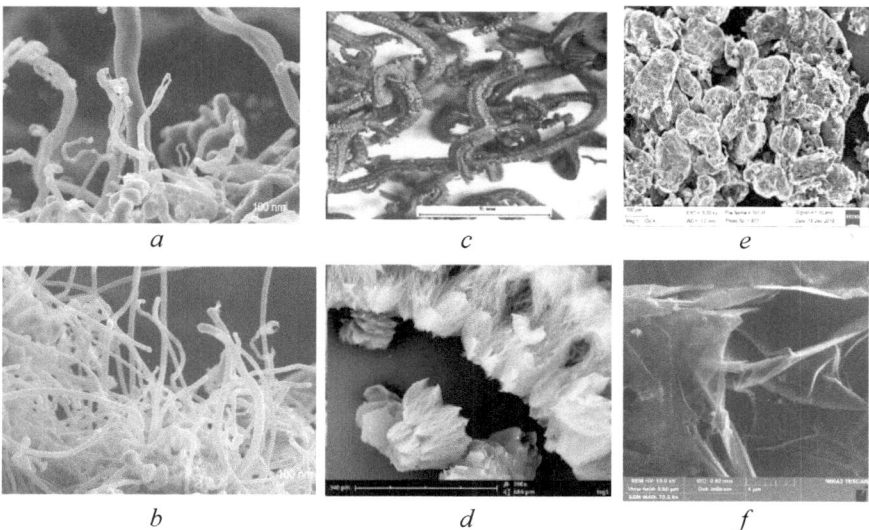

Fig. 3.46 Microstructure of carbon nanomaterials: **a, b**—multi-walled carbon nanotubes; **c, d**—thermally expanded graphite; **e, f**—globular multilayer graphene

of determining the wave number was in the range of 0.5–2 cm^{-1}. The exposure time was 100 ms, and 50 spectra were averaged. In each case, the Raman spectra of all samples were recorded using the same parameters. The Raman spectra confirmed the presence of carbon nanotubes with two characteristic peaks at 1300 and 1590 cm^{-1}, which correspond in Fig. 3.47 to D and G modes, respectively. The first, second, and fourth spectra (Fig. 3.47) are typical spectra of nanotubes with characteristic D and G modes. In all three samples, the D mode had a higher intensity than the second one, which is not typical for intact single-walled nanotubes. Therefore, one can conclude that the nanotubes deposited by the CVD method were multi-walled and most likely had a curved shape. In all three samples, a shoulder in the G mode was clearly observed (which may be an overtone of the D mode) that indicated the high density of defects in the samples.

In order to determine the dependence of density and thermal conductivity on pressing pressure in the range of 170–450 MPa, sets of samples from carbon nanotubes, thermally expanded graphite, and globular multilayer graphene were produced, Fig. 3.48.

Thermally expanded graphite is a product of multistage technological processing [69] of natural crystalline graphite powder. Figure 3.26c shows the appearance of TEG particles at the macrolevel, which consist of individual worm-shaped 5–10-mm long particles. The microstructure of TEG at the microlevel is shown in Fig. 3.46d: the structure of a single particle is seen, which consists of separate graphene clusters comprised of 100–500 elements. According to [70], the main stages of TEG production include oxidation of flaky graphite, its drying, thermal expansion, and

Fig. 3.47 Raman spectra of carbon nanomaterials and images of synthesized carbon nanotubes: **a** Raman spectra of CNTs; **b** CNTs on an iron catalyst; **c** CNT samples in glass tubes for determination of Raman spectra

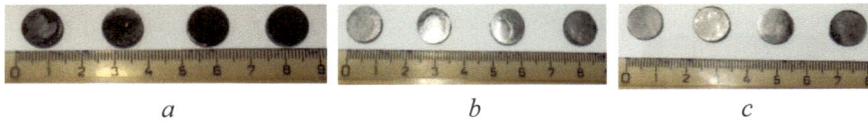

Fig. 3.48 Appearance and dimensions of: **a** carbon nanotubes; **b** thermally expanded graphite; **c** globular multilayer graphene

subsequent processing aimed at obtaining the final product or part. The mechanism of transformation of oxidized graphite into TEG was also analyzed; it is related to a sharp increase in pressure between the layers due to rapid heating (thermal shock). The pressure that occurs during the thermal shock causes foaming and intermolecular explosion, which are accompanied by the formation of a peculiar foam-like structure [70].

In [71], a method of producing globular multilayer graphene was described. The technique includes mechanical processing of TEG with the aim of destruction it into separate graphene layers (from several to dozens of layers) and forming globules with approximately the same size (~100 µm). The microstructure of individual globules is shown in Fig. 3.46e, and the morphology of globules comprised of crumpled graphene layers is seen in Fig. 3.46f.

We have developed a method for producing globular graphite, which consists of several stages. The first stage is processing of TEG in a cavitation apparatus with the aim of its temporal transformation from a hydrophobic material into a hydrophilic one. The second stage is the rapid ultrasound processing of TEG soaked in a liquid—the destruction of TEG into individual graphene layers mainly occurs at this stage. The third stage is drying followed by finishing in a mill with the aim to form the structure of the material (in particular, to crush coarse globules and form spherical globules of the same size). A detailed description of the equipment and techniques for producing TEG is given in [72].

The thermophysical parameters of thermally expanded graphite, carbon nanotubes, and globular multilayer graphene were determined by the pulse method for measuring thermal conductivity (PMMTC). This is a non-destructive and non-contact method. In this method, the front surface of the sample is irradiated with a pulse of light in the visible range ($\tau = 100$ ms). This is the "primary" radiation flow. The absorbed part of this "primary" flow is converted into thermal energy, which causes an increase in temperature and gives rise to the propagation of temperature waves into the sample. A certain part of not dissipated heat flow causes an increase in temperature on the opposite back surface of the sample, where it converts into a "secondary" flow of infrared (IR) radiation. This "secondary" flow of IR radiation is delayed relative to the incident "primary" one due to the finite rate of the heat diffusion process; it is registered by an IR detector. When a pyroelectric sensor [73] is used as the IR detector, there is no need to know the values of the incident and absorbed energy, the absorption coefficient of the front surface, and the emission coefficient of the back surface of the sample, as well as the profile of the temperature increase over time and the pyroelectric parameters of the sensor [73].

The block diagram of the PMMTC measuring system is shown in Fig. 3.49. A light-emitting diode (LED) as a radiation heater and a pyroelectric detector as a thermal sensor (PyES) were used. The delay time between "primary" heat flow (without a sample) at PyES and "secondary" heat flow (with a sample) is illustrated in Fig. 3.49b-d.

Figure 3.49 shows the block diagram of the implementation of the thermal conductivity measurements by the pulse method for carbon nanomaterials.

For the impulse method of measuring thermal conductivity, we will use the equation of thermal conductivity $\frac{C\rho d Q(r,t)}{dt} = k\Delta Q(r,t) + W(r,t)$ with the following boundary conditions:

$$-\frac{k\partial Q}{\partial x}\Big|_0 = \alpha_1 Q(0,t), \quad -\frac{k\partial Q}{\partial x}\Big|_1 = \alpha_2 Q(L,t), \tag{3.5}$$

where $W(r,t)$ is the thermal impulse shape, $C\rho$ is the specific heat capacity per unit volume, k is the thermal conductivity coefficient (W/(m K), L is the layer thickness, and α_1 and α_2 are the layer heat loss coefficients.

The thermal boundary problem consists in determining the radiation density of the radiation on the back surface of the layer that was pulse irradiated [74]:

$$q_\rho(t) = q_{\rho 0}\{\eta(t) - \eta(t - \tau_U)\},$$

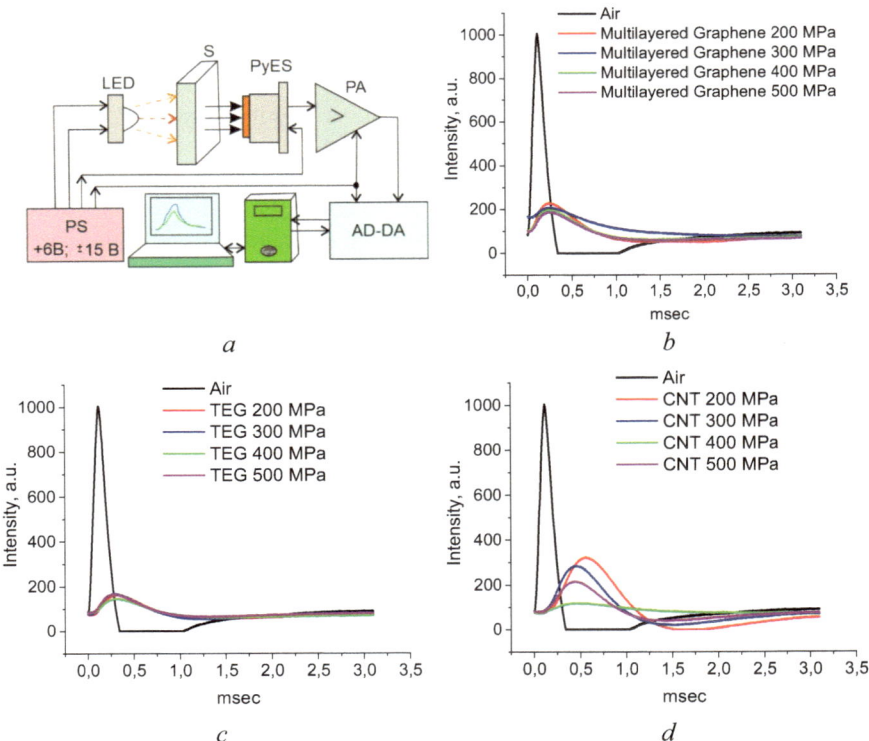

Fig. 3.49 Block diagram of the pulse device for measuring thermal conductivity [73]: **a** LED
light diode (GREE-XML), S—a sample from carbon materials, PyEs—pyroelectric sensor (IRA-
E700ST0), PA—amplifier (Ku = 10,000), AD-DA—ADC-DAC block (± 10 V, 12-digit, 2 channels), PS—power supply (± 15 V, + 6 V), PC—personal computer, **b–d** delay time between the
"primary" and "secondary" heat flows in globular multilayer graphene, thermally expanded graphite,
and carbon nanotubes, respectively

$$\rho C \frac{\partial T}{\partial t} = \frac{\partial}{\partial z}\left(k \frac{\partial T}{\partial z}\right).$$

The temperature distribution during implementation of the pulse method of
measuring thermal conductivity is schematically shown in Fig. 3.50.

The amount of heat that passed through a sample is calculated using the formula
$W\rho^{(x=l)} = \alpha\{Q(L, t) - Q(L, t - \tau_U)\}$. The boundary conditions are specified in
(5). The solution has the following form: $\tau_d = \frac{L^2}{\chi} = \frac{C\rho L^2}{k}$; $\chi = \frac{k}{C\rho}$, where χ is the
coefficient of thermal conductivity [m²/s], τ_d is the time shift of the maximum of
the passed thermal pulse, C is the isobaric heat capacity, and ρ is the layer density.

The densities and thicknesses of the samples for calculating the thermal conductivity values of carbon nanotubes, thermally expanded graphite, and globular
multilayer graphene are listed in Table 3.4.

Fig. 3.50 Model of the heat flow redistribution through a sample (thermally expanded graphite, carbon nanotubes, or globular multilayer graphene) during PMMTC

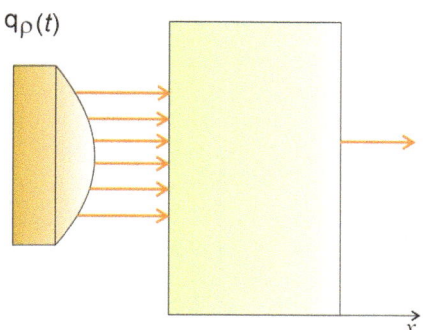

$q_\rho(t)$

Table 3.4 Calculation of thermal conductivity

Sample number	Sample mass, g	Volume, cm^3	Density, g/cm^3	Sample thickness, [10^{-6} m]. L, m	Air, τ, [10^{-3} s]	Sample, τ, [10^{-3} s]	χ, [m^2/s] 10^{-6} Diffusivity
CNT							
1	0.240	0.258	0.99	2280	120	557	1.2
2	0.230	0.171	1.34	1510	120	457	6.76
3	0.240	0.181	1.32	1600	120	469	7.3
4	0.250	0.184	1.36	1630	120	445	8.17
Graphene							
1	0.110	0.062	1.77	550	120	250	2.32
2	0.130	0.068	1.90	680	120	250	2.36
3	0.150	0.071	2.11	710	120	256	2.92
4	0.160	0.078	2.05	780	120	256	3.50
TEG							
1	0.170	0.188	0.90	1660	120	315	1.41
2	0.150	0.115	1.30	1020	120	280	6.50
3	0.160	0.107	1.50	1030	120	280	6.65
4	0.190	0.118	1.61	1040	120	280	6.7

All studied materials (CNT, TEG, and GMLG) consisted of the same substance, carbon; however, they had different methods of synthesis and structures at the macro and microlevels. The dependences of the density of CNMs on the pressing pressure are shown in Fig. 3.51. As can be seen, increasing the pressing pressure in the range of 170–450 MPa leads to a slight increase in the density of CNMs. Under the given pressing conditions, samples had density 0.9–2.05 g/cm^3. The dependence of the density of TEG on the pressing pressure shows that this material can be compacted without deformation and change of shape. Nanotubes and globular multi-layer graphene partially recovered their initial shape after the load was removed due to elastic aftereffect.

The thermophysical properties of carbon nanomaterials of the same brand can vary significantly. This is explained by different porosity of the materials, which has a significant effect on their thermal conductivity. The porosity of carbon materials of the same brand is uniquely related to the macroscopic density of the samples. It is found out that the density of GMLG, CNT, and TEG increases with pressing pressure increasing, see Fig. 3.51a. In the process of TEG compaction, it was found that TEG powders had an extended zone of structural compaction; i.e., the material formed in a wide range (Fig. 3.51a) of densities at pressures 170–450 MPa. It was found out that, at the first stage of the formation process, at low loads, the powder particles structurally repacked, then compaction occurred due to the extended surface of the particles (which had a popcorn-like shape), and finally they deformed. However, the

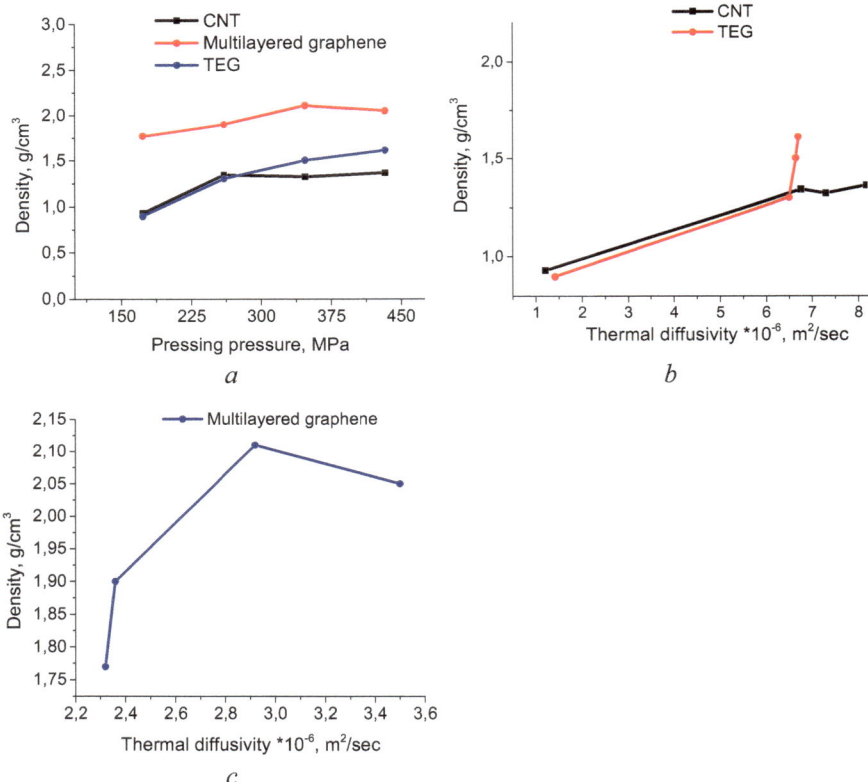

Fig. 3.51 Dependences of density of carbon nanomaterials on pressing pressure at room temperature 293 K: **a** carbon nanotubes, thermally expanded graphite, globular multilayer graphene; **b, c** on thermal conductivity

compaction of CNT and GMLG occurred slowly in a small range without significant changes. It is shown that the CNTs and GMLGs during formation at pressing pressures 200–500 MPa without binding additives begin to delaminate and lose strength. It is found out that the thermal conductivity of GMLG, CNT, and TEG increases with increasing density and pressure (Fig. 3.51b,c). The mechanism of changes in thermal conductivity of these systems is likely related to electron-lattice interaction. Heat transfer can occur by various transport mechanisms (collisions, diffusion). However, exactly the electronic mechanism can be responsible for the increase in thermal conductivity. The relationship between the thermal conductivity of carbon nanomaterials and their macroscopic density is determined.

Summary

The structural features of thermally expanded graphite, carbon nanotubes, and globular multilayer graphene, which were produced at the equipment of Gas Institute of N.A.S. of Ukraine, have been studied.

A device for the pulse method of measuring thermal conductivity of diverse materials with registration of information on a PC has been developed. This system was developed at the Technical Center, N.A.S. of Ukraine. Thermal conductivity of carbon nanomaterial samples produced at different pressing pressures was measured. It is determined that the thermal conductivity of CNMs increases with increasing density and pressure. It is shown that the mechanisms of heat transfer in these systems are related to the electron–phonon interaction. Heat transfer can occur by an electronic or diffusion mechanism; nevertheless, exactly the electronic mechanism causes the increase in thermal conductivity of the samples of thermally expanded graphite, carbon nanotubes and globular multilayer graphene.

References

1. Boehm H, Setton R, Stumpp E (1994) Nomenclature and terminology of graphite intercalation compounds (IUPAC recommendations 1994). Pure Appl Chem 66(9):1893–1901. https://doi.org/10.1351/pac199466091893
2. Geim A, Novoselov K (2007) The rise of grapheme. Nat Mater 6(3):183–191. https://doi.org/10.1038/nmat1849
3. Rao C, Sood A, Subrahmanyam K, Govindaraj A (2009) Graphene: the new two-dimensional nanomaterial. Angew Chem Int Ed Engl 48(42):7752–7777. https://doi.org/10.1002/anie.200901678
4. Rao C, Sood A, Voggu R, Subrahmanyam K (2010) Some novel attributes of graphene. Chem Lett 1:572. https://doi.org/10.1021/jz9004174
5. Jang B, Zhamu A (2008) Processing of nanographene platelets (NGPs) and NGP nanocomposites: a review. J Mater Sci 43(15):5092–5101. https://doi.org/10.1007/s10853-008-2755-2
6. Murugan A, Muraliganth T, Manthiram A (2009) Rapid, facile microwavesolvothermal synthesis of graphene nanosheets and their polyaniline nanocomposites for energy strorage. Chem Mater 21(21):5004–5006. https://doi.org/10.1021/cm902413c
7. Stoller M, Park S, Zhu Y, An J, Ruoff R (2008) Graphene-based ultracapacitors. Nano Lett 8:3498–3502. https://doi.org/10.1021/nl802558y

8. Soldano C, Mahmood A, Dujardin E (2010) Production, properties and potential of graphene. Carbon 48:2127. https://doi.org/10.1016/j.carbon.2010.01.058
9. Zhi L, Mullen K (2008) A bottom-up approach from molecular nanographenes to unconventional carbon materials. J Mater Chem 18:1472. https://doi.org/10.1039/B7175 85J
10. Loh K, Bao Q, Ang P, Yang J (2010) The chemistry of graphene. J Mater Chem 20(12):2277–2289. https://doi.org/10.1039/b920539j
11. Birkhoff G, Zarantonello EH (1957) Jets, wakes, and cavities. Academic, New York. https://doi.org/10.1017/S0022112058210100
12. Strativnov E, Bondarenko B, Dmitriev V, Kozhan A, Khovavko A (2021) Method of graphene producing. Patent 149271 Ukr.,C01B32/182 (Ukr.)
13. Svishchev G (1994) Venturi tube. Aviation: encyclopedia. Great Russian Encyclopedia, Moscow
14. Yutkin L (1986) Electrohydraulic effect and its application in industry. Mashinostroenie, Leningrad Zhi L, Mullen K (2008) A bottom-up approach from molecular nanographenes to unconventional carbon materials. J Mater Chem 18:1472. https://doi.org/10.1039/B7175 85J
15. Strativnov E, Kozhan A, Bondarenko B (2011) Method of obtaining thermally expanded graphite. Patent. 99875 Ukr., C01B 31/04 (Ukr.)
16. Kucinskis G, Bajars G, Kleperis J (2013) Graphene in lithium ion battery cathode materials: a review. J Power Sources 240:66–79. https://doi.org/10.1016/j.jpowsour.2013.03.160
17. Tsang CHA, Huang H, Xuan J, Wang H, Leung DYC (2020) Graphene materials in green energy applications: recent development and future perspective. Renew Sustain Energy Rev 120:109656. https://doi.org/10.1016/j.rser.2019.109656
18. Dhinakaran V, Stalin B, Swapna Sai M, Vairamuthu J, Marichamy S (2021) Recent developments of graphene composites for energy storage devices. Mater Today Proc 45:1779–1782. https://doi.org/10.1016/j.matpr.2020.08.631
19. Olabi AG, Abdelkareem MA, Wilberforce T, Sayed ET (2021) Application of graphene in energy storage device—a review. Renew Sustain Energy Rev 135:110026. https://doi.org/10.1016/j.rser.2020.110026
20. Kirillov SA (2019) Electrode materials and electrolytes for high-rate electrochemical energy systems: a review. Theor Exp Chem 55:73–95. https://doi.org/10.1007/s11237-019-09598-2
21. Potapenko AV, Kirillov SA (2014) Lithium manganese spinel materials for high-rate electrochemicalapplications. J Energy Chem 23:543–558. https://doi.org/10.1016/S2095-495 6(14)60184-4
22. Strativnov E, Khovavko A, Guochao N (2022) Obtaining of globular graphene based on thermally expanded graphite. Appl Nanosci 12:2791–2811. https://doi.org/10.1007/s13204-022-02589-1
23. Strativnov EV (2015) Design of modern reactors for synthesis of thermally expanded graphite. Nanoscale Res Lett 10:244. https://doi.org/10.1186/s11671-015-0919-y
24. Potapenko O, Potapenko A, Zhou C, Zhang L, Xu J, Gu Z (2020) Improved effect of water-soluble binder NV-1A on theelectrochemical proprieties ofLFP electrodes. Russ J Electrochem 56:1043–1050. https://doi.org/10.1134/S1023193520120174
25. Bianco A, Cheng HM, Enoki T, Gogotsi Y, Hurt RH, Koratkar N, Kyotani T, Monthioux M, Park CR, Tascon JMD, Zhang J (2013) All in the graphene family—a recommended nomenclature fortwodimensional carbon materials. Carbon 65:1–6. https://doi.org/10.1016/j.carbon.2013.08.038
26. Ferrari AC, Meyer JC, Scardaci V, Casiraghi C, Lazzeri M, Mauri F, Piscanec S, Jiang D, Novoselov KS, Roth S, Geim AK (2006) Raman spectrum of graphene and graphene layers. Phys Rev Lett 97:187401. https://doi.org/10.1103/PhysRevLett.97.187401
27. Yin P, Lin Q, Duan Y (2020) Applications of Raman spectroscopy in two-dimensional materials. J Innov Opt Health Sci 13:2030010. https://doi.org/10.1142/S1793545820300104
28. Vargas COA, Caballero A, Morales J (2012) Can the performance of graphene nanosheets for lithium storage in Li-ionbatteries be predicted? Nanoscale 4:2083–2092. https://doi.org/10.1039/c2nr11936f

29. Han S, Wu D, Li S, Zhan F, Feng X (2013) Graphene: a two-dimensional platform for lithium storage. Small 9:1173–1187. https://doi.org/10.1002/smll.201203155

30. Qi X, Wang F, Xie H, Mao L, Mao J (2021) Ultra-high capacity dualion batteries realized by few-layered reduced graphene oxide and cathode structuredesign. J Mater Sci 56:10555–10564. https://doi.org/10.1007/s10853-021-05938-7

31. Konjushkov AG (2007) Experimental study sites seals reactor plants with VVER. Ph.D. (Tech.). Podolsk [in Russian]

32. Ryzhov SB (2007) Development, calculation and experimental justification and pilot-industrial operation of VVER-1000 reactor compaction units. Ph.D. (Tech.), St. Petersburg [in Russian]

33. Rus'janov VG, Denisov VP, Dragunov YuG, Seleznev AV, Ryzhov SB, Geront'ev AE, Konjushkov AG (2004) Sealing device plug connections of equipment of reactor plants VVER. Moskva [in Russian]

34. Seleznev AV, Geront'ev AE, Konjushkov AG, Ryzhov SB (2002) Development and introduction of seal assem blies with expanded graphite gaskets in the equipment of reactor units wwer. Annual report of FSUE OKB "Gidropress" for 2001. Main research and development work. Podolsk [in Russian]

35. Geront'ev AE, Strahov AA, Konjushkov AG, Alekseev DE (2005) Modernizacija uplot-nitel'nyh ustrojstv parogeneratorov PGV-440 i PGV1000 M s primeneniem prokladok iz rasshirennogo grafita. Voprosy atomnoj nauki i tehniki 9:95–101 [in Russian]

36. Geront'ev AE, Strahov AA, Konjushkov AG, Alekseev DE (2005) Modernizacija uplot-nitel'nyh ustrojstv parogeneratorov AJeS s VVJeR-440, VVJeR-1000. Atomnaja jenergija 6:476–481 [in russian]

37. Ryzhov SB, Konjushkov AG, Titov OV (2005) Razrabotka, raschjotno-jeksperimental'noe obosnovanie i opytnopromyshlennaja jekspluatacija uzlov uplotnenij reaktorov tipa VVJeR-1000. Voprosy atomnoj nauki i tehniki 9:103–115 [in Russian]

38. Konjushkov AG, Rus'janov VG, Geront'ev AE (2004) Obshhie voprosy uplotnenij raz'emnyh soedinenij RU s VVJjeR. Sbornik trudov FGUP OKB «Gidropress» 5(Part 2):287–293 [in Russian]

39. Seleznev AV, Geront'ev AE, Konjushkov AG (2001) Razrabotka i vnedrenie uzlov uplotnenij s prokladkami iz rasshirennogo grafita v oborudovanii reaktornyh ustanovok vver. Sbornik trudov FGUP OKB «Gidropress» 2 (Part 3):435–439 [in Russian]

40. A.s. 1266103 SSSR. mKI4 C01B34/04. Sposob poluchenija rasshirennogo grafita. B.E. Paton, A.P. Kozhan, V.K. Pikalov, K.E. Mahorin. No. 8513455/43; zajavl. 12.05.85; opubl. 22.06.86, bjul. No. 14 [in Russian]

41. Mahorin KE, Kozhan AP (1987) Vspuchivanie grafita v plotnom i vzveshennom slojah. Himicheskaja tehnologija 2:14–19 [in Russian]

42. Patent of Ukraine N 115288. Bondarenko B.I., Kozhan A. P., Dmitriev V.M., Kul'chic'kij G.M., Rjabchuk V.S., Pisarenko I.O., Chernjuk I.M. Method of manufacturing of sealing elements from thermoexpanded graphite [in Ukrainian]

43. Kormak D (1989) Struggle against pollution of the sea by oil and chemical substances. Transport 365

44. Strativnov EV, Kozhan AP, Dmitriev VM, Sergienko AA, Vavrish AS (2012) A study of regeneration process of thermal expanded graphite. Energotechnologii i resursozberezhenie. 1:47–52

45. Pat. 104098 Ukr., IASC (2012) B09C1/00. Method of cleaning water and soil from oil and oil products by graphite sorbent. / Dmitriev V.M., Kozhan A.P., Ryabchuk V.S., Strativnov E.V., Bondarenko O.B., – Publ. 12.08.2013

46. Pat. 79769 Ukr., IASC (2012) B09C1/00. Method of cleaning water and soil from oil and oil products oil-absorbing by sorbent based on expanded graphite. / Dmitriev V. M., Kozhan A. P., Bondarenko O. B., Strativnov E.V., Ryabchuk V.S., Pisarenko A.I. Publ. 25.04.2013

47. Pat. 85362 Ukraini, MPK7 B01J20/00. Portable autonomous apparatus in order to receive and application of sorbent. /Bondarenko B.I., Kozhan O.P., Dmitriev V.M., Raybchuk V.S., Sergienko O.A. Opubl. 12.10.2009, Byul.№1

48. Method of granulation of carbon sorbent: Pat. 23972 Ukraini, MPK7B 01J20/00, B09J20/20/ Bondarenko B.I., Kozhan O.P., Dmitriev V.M., Sergienko O.A., Moskalik L.D. (Ukraini)—Zaaivnikivlasnik: Institutgazu NAN Ukraini. −№ u200701849; zayavl.22.02.2007; opubl.11.06.07, Byul.№ 8.-8 s
49. Method of granulation of oil absorbing sorbent on the basis termoexpanded graphite: Pat. 31784 Ukraini, МПК7 B01J20/00, B09J20/20/ Bondarenko B.I., Kozhan O.P., Dmitriev V.M., Hohulay I.M., Raybchuk V.S. (Ukraini) – Zayaivnikivlasnik: Institutgazu NAN Ukraini. −№ u200713038; zaaivl.26.11.2007; opubl.25.04.08, Byul.№ 8.-7c
50. Kozhan AP, Dmitriev VM, Strativnov EV, Ryabchuk VS, Bondarenko OB (2012) Cleaning surface of water basins and soil at oil outflow with using a sorbent on thermoexpanded graphite basis. Ecologiya i promishlennost. 4:33–42
51. Method of water cleaning from oil and oil products with the help of carbon sorbent: Pat. 31783 Ukraini, МПК7 E02B15/04, C02F1/28/ Bondarenko B.I., Kozhan O.P., Dmitriev V.M., Hohulay I.M., Aleksandrov V.V. (Ukraini)—Zayaivnikivlasnik: Institut gazu NAN Ukraini. -№ u200713037; zaaivl.26.11.2007; opubl.25.04.08, Byul.№ 8.-7c
52. Cleaning method of water surface from hydrophobic pollution: Pat. 82819 Ukraini, МПК7 E02B15/04, C02F1/28/Bondarenko B.I., Kozhan O.P., Dmitriev V.M., Hohulay I.M., Aleksandrov V.V. (Ukraini)—Zayavnikivlasnik: Institutgazu NAN Ukraini. −№ u200713037; zayvl.26.11.2007; opubl.25.04.08, Byul.№ 8.- 7c
53. Kozhan AP, Dmitriev VM, Strativnov EV, Ryabchuk VS, Bondarenko OB (2012) Purification of water surface and soil from accidental spills of oil and oil products by sorbent based on thermally expanded graphite. In: Innovative solutions of actual problems of basic industries, environment, energy and resources: proceedings of the international scientific-practical conference (Kazantip–ECO–2012), Shchelkino, Crimea/Ukrgt "Energostal", 4–8 June 2012, in 3 volumes. NTMT, Kharkov, vol 3, pp 73–84
54. Kozhan AP, Dmitriev VM, Strativnov EV, Ryabchuk VS, Bondarenko OB (2012) Oil-absorbing sorbent based on nano-layered thin films of graphite. In: 3rd international scientific conference on Nanostructured Materials 2012: Russia–Ukraine–Belarus: proceedings of the international scientific-practical conference (NANO-2012), Saint-Petersburg. Russia Scientific center of Russian Academy of Sciences (SPBSC RAS), Institute of silicate chemistry of the. I.V. Grebenshchikov, 19–22 Nov 2012, p 274
55. Method of cleaning of soil and coastal strip from hydrophobic pollution: Pat. 83181 Ukraini, МПК7 B 09 C1/00, B01J20/20 /Bondarenko B.I., Kozhan O.P., Dmitriev V.M., Raybchuk V.S., Komisarenko A.A. (Ukraini)—Zayavnik i vlasnik: Institut gazu NAN Ukraini. −№ u200701852; zayavl. 22.02.2007; opubl. 25.06.08, Byul. № 12.- 6 c
56. Cleaning method of water surface from oil and hydrophobic liquid: Pat. 92000506 Rossii, МПК7 C01B31/04 Smirnov A.V. (Rossia)—Zayavitel i sobstvennik Smirnov A.V. −№ 93000506/26; zayavl 15.10.1993; opubl. 27.06.96, Byul.№ 11. – 5 c
57. Cleaning method of water surface from oil and oil products: Pat. 2140488 Rossii, МПК7 Э02B15/04/Samosadnii V.P. (Rossiya)—Zayavitel i sobstvennik: OOO Nauchno-proizvodstvennoe obedinenie" Tehnoprom" - № 98001306/32; zayavl 21.12.1998; opubl.27.10.99, Byul. № 7.– 6c
58. Cleaning method of soil from oil and oil products by graphite sorbent: Pat. 48026 Ukraini, МПК9 У 09 3 1/00 /Bondarenko B.I., Kozhan O.P., Dmitriev V.M., Raybchuk V.S., Sergienko O.A., (Ukraini)—Zayavnik i vlasnik: Institut gazu NAN Ukraini. -№ u200811085; zaaivl12.09.2008.;opubl. 10.03.10, Byul.№ 5.– 6 c
59. Cleaning method of soil from oil products sorbent graphite: Pat. 83180 Ukraini, МПК7 B 09C1/00, C02F1/28/ Bondarenko B.I., Kozhan O.P., Dmitriev V.M., Raybchuk V.S., Sergienko O.A., (Ukraini) – Zayavnik i vlasnik: Institut gazu NAN Ukraini. -№ u200701850; zayavl 22.02.2007.; opubl.25.06.08, Byul.№ 12.– 9 c. 33
60. Vorob'eva AI (2010) Equipment and techniques for carbon nanotube research. Usp Fiz Nauk 180(3):265–288. https://doi.org/10.3367/UFNr.0180.201003d.0265
61. Barabash MYu, Suprun NP, Kolesnichenko AA, Leonov DS, Lytvyn RV (2020) Control and Amplification effects of raman scattering on amorphous carbon films with a copper sublayer

with the participation of iron phthalocyanine. Nanosyst Nanomater Nanotechnol 18(1):41–52. https://doi.org/10.15407/nnn.18.01

62. Stankus CV, Savchenko IV, Agazhov AS, Yatsuk OS, Zhmurikov EI (2013) Thermophysical properties of graphite MPG6. Thermophys High Temp 51(2):205–209

63. Landau LD (1986) Theoretical physics, vol 10. Landau LD, Livshits EM. Hydrodynamics, vol. 6. Nauka, Moscow, 736p

64. Karslow G (1964) Thermal conductivity of solids. Karslow G Eger. M., Nauka, 489p

65. Shpak AP, Grechko LG, Kunitska LYu, Lerman LB, Semchuk OYu (2007) Periodic structures induced on the surface of solid bodies by the interference of laser beams. Thermal Effects Nanosyst Nanomater Nanotechnol 5(3):683–718

66. Nebesnyi A, Kotov V, Sviatenko A, Filonenko D, KhovavkoA, Bondarenko B(2017) Carbon nanomaterial formationon fresh-reduced iron by converted natural gas. Nanoscale Res Lett 12:107. https://doi.org/10.1186/s11671-017-1882-6

67. Nebesniy A, Kotov, V, Svyatenko M, Filonenko D, Khovavko A, BondarenkoB (2015) Formation of carbon nanomaterial during treatment of freshly reduced iron with converted natural gas. Energy Technol Resour Conserv 5–6:34–42

68. Khovavko A, Filonenko D, Barabash M, Nebesnyi A, Sviatenko A, Trosnikova I (2023) NieGMulti-walled carbon nanotubes synthesis on iron ore pellets by CVD method. Appl Nanosci. https://doi.org/10.1007/s13204-023-02954-8

69. Pat. 99875 Ukr, MPK (2011) C01B 31/04. Method for producing thermally expanded graphite Strativnoav E.V., Kozhan A.P., Bondarenko B.I. Published. 10.10.2012.

70. Strativnov EV (2015) Design of modern reactors for synthesis of thermally expanded graphite. Nanoscale Res Lett 10:245. https://doi.org/10.1186/s11671-015-0919-y

71. Strativnov EV, Bondarenko BI, Dmitriyev VM, Kozhan AP, Khovavko AI (2021) Pat. 149271 Ukr, MPK (2021) C01B32/182. Method forproducing graphene, 04 Nov 2021

72. Strativnov E, Guochao N, Khovavko A (2022) Obtaining of globular graphene based on thermally expanded graphite. Appl Nanosci. https://doi.org/10.1007/s13204-022-02589-1

73. Morozovsky NV, Barabash YuM, Grebelna YuV, Kartel MT (2023) Sementsov YuI, Dovbeshko GINonstationary thermophysical characterization of exfoliated graphite with carbon nanotubes composites. Low Temp Phys 49(5):604–620. https://doi.org/10.1063/10.0017817

74. Sun JG (2010) Thermal conductivity measurements for thermal barrier coatings based on one and two-sided thermal imaging methods. Review of progress in quantitative nondestructive evaluation. In: AIP Conference proceedings 1211, vol 29, p 458. https://doi.org/10.1063/1.3362429

75. Taylor RE, Maglic KD (1984) Pulse method for thermal diffusivity measurement. In: Maglic KD et al (eds) Compendium of Thermophysical Property Measurement Methods (1) Survey of Measurement Techniques. Plenum Press, New York, pp 305–336